Praxis Algebra I (5162) Made Easy

Ultimate Study Guide and Test Prep with Key Points, Examples, and Practices. The Best Praxis Algebra I (5162) Tutor for Beginners and Advanced Students + Two Full-Length Practice Tests

Dr. Abolfazl Nazari

Copyright © 2024 Dr. Abolfazl Nazari

PUBLISHED BY EFFORTLESS MATH EDUCATION

EFFORTLESSMATH.COM

All rights reserved. No part of this publication may be reproduced, distributed, or transmitted in any form or by any means, including photocopying, recording, or other electronic or mechanical methods, without the prior written permission of the author, except in the case of brief quotations embodied in critical reviews and certain other noncommercial uses permitted by copyright law, including Section 107 or 108 of the 1976 United States Copyright Act.

Copyright ©2024

Welcome to Praxis Algebra I (5162) Made Easy 2024

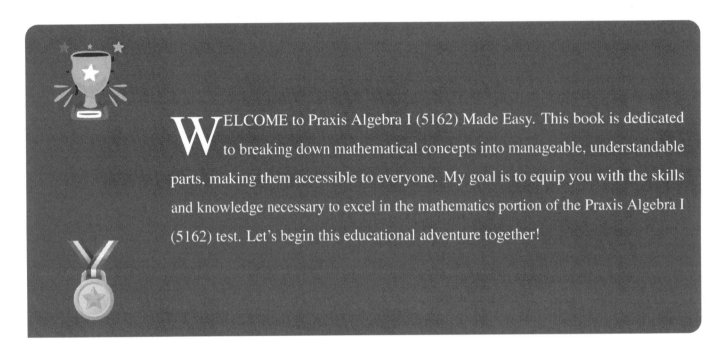

WELCOME to Praxis Algebra I (5162) Made Easy. This book is dedicated to breaking down mathematical concepts into manageable, understandable parts, making them accessible to everyone. My goal is to equip you with the skills and knowledge necessary to excel in the mathematics portion of the Praxis Algebra I (5162) test. Let's begin this educational adventure together!

Praxis Algebra I (5162) Made Easy provides comprehensive coverage of the key mathematical topics needed for the Praxis Algebra I (5162) test. The book is structured into detailed chapters on Fractions and Mixed Numbers, Decimals, Integers and Order of Operations, Ratios and Proportions, Percentages, and much more. Each chapter starts with basic concepts and gradually moves to more complex ones, ensuring you gain a complete understanding of each topic. The content is tailored to not only prepare you for the Praxis Algebra I (5162) test but also to apply these skills in real-life situations.

In keeping with the *Math Made Easy* series' philosophy, this book adopts an interactive and practice-oriented approach to learning. Each mathematical concept is introduced in a clear and straightforward manner, accompanied by examples to help illustrate its application. A variety of practice problems are provided to mirror the style and challenges of the Praxis Algebra I (5162) test, enabling you to test your knowledge and strengthen your understanding. I am excited to show you what the book contains.

What is included in this book

- ☑ Online resources for additional practice and support.
- ☑ A guide on how to use this book effectively.
- ☑ All Praxis Algebra I (5162) concepts and topics you will be tested on.
- ☑ End of chapter exercises to help you develop the basic math skills.
- ☑ Praxis Algebra I (5162) test tips and strategies.
- ☑ 2 realistic and full-length practice tests with detailed answers.

Effortless Math's Praxis Algebra I Online Center

Effortless Math Online Praxis Algebra I Center offers a complete study program, including the following:

- ☑ *Step-by-step instructions on how to prepare for the Praxis Algebra I (5162) test*
- ☑ *Numerous Praxis Algebra I (5162) worksheets to help you measure your math skills*
- ☑ *Complete list of Praxis Algebra I (5162) formulas*
- ☑ *Video lessons for all Praxis Algebra I (5162) topics*
- ☑ *Full-length Praxis Algebra I (5162) practice tests*

Visit `EffortlessMath.com/PraxisAlgebra1` to find your online Praxis Algebra I (5162) resources.

Scan this QR code

(No Registration Required)

Tips for Making the Most of This Book

This book is all about making mathematics easy and approachable for you. Our aim is to cover everything you need to know, keeping it as straightforward as possible. Here is a guide on how to use this book effectively: First, each math topic has a core idea or concept. It's important to understand and remember this. That's why we have highlighted key points in every topic. These are like mini-summaries of the most important stuff.

Examples are super helpful in showing how these concepts work in real problems. In every topic, we've included a couple of examples. If you feel very smart, you can try to solve them on your own first. But they are meant to be part of the teaching; They show how key concepts are applied to the problems. The main thing is to learn from these examples.

And, of course, practice is key. At the end of each chapter, you will find problems to solve. This is where you can really sharpen your skills.

To wrap it up:

- *Key Points*: Don't miss the key points. They boil down the big ideas.
- *Examples*: Try out the examples. They show you how to apply what you're learning.
- *Practices*: Dive into the practice problems. They're your chance to really get it.

In addition to the material covered in this book, it is crucial to have a solid plan for your test preparation. Effective test preparation goes beyond understanding concepts; it involves strategic study planning and practice under exam conditions.

- **Begin Early.** Start studying well before the exam to avoid rushing, allowing for a thorough review.
- **Daily Study Sessions.** Study regularly for 30 to 45 minutes each day to enhance retention and reduce stress.
- **Active Note-Taking.** Write down key points to internalize concepts and improve focus. Review notes regularly.
- **Review Challenges.** Spend extra time on difficult topics for better understanding and performance.
- **Practice.** Engage in extensive practice using end-of-chapter problems and additional workbooks.

Explore other guides, workbooks, and tests in the series to complement your study, offering extra practice and enhancing understanding, problem-solving skills, and academic preparation.

Contents

1	**Fundamental and Building Blocks**	1
1.1	Adding and Subtracting Integers	1
1.2	Multiplying and Dividing Integers	2
1.3	Order of Operations	2
1.4	The Distributive Property	3
1.5	Integers and Absolute Value	4
1.6	Proportional Ratios	4
1.7	Similarity and Ratios	5
1.8	Percent Problems	6
1.9	Percent of Increase and Decrease	7
1.10	Discount, Tax, and Tip	9
1.11	Simple Interest	10
1.12	Approximating Irrational Numbers	11
1.13	Practices	12
2	**Exponents and Variables**	19
2.1	Mastering Exponent Multiplication	19

2.2	Exploring Powers of Products	19
2.3	Mastering Exponent Division	20
2.4	Understanding Zero and Negative Exponents	21
2.5	Working with Negative Bases	21
2.6	Introduction to Scientific Notation	22
2.7	Addition and Subtraction in Scientific Notation	23
2.8	Multiplication and Division in Scientific Notation	23
2.9	Practices	24

3	**Expressions and Equations**	**31**
3.1	Translate a Phrase into an Algebraic Statement	31
3.2	Simplifying Variable Expressions	32
3.3	Evaluating Single Variable Expressions	32
3.4	Evaluating Two Variable Expressions	33
3.5	Solving One-Step Equations	34
3.6	Solving Multi–Step Equations	35
3.7	Rearranging Equations with Multiple Variables	35
3.8	Finding Midpoint	36
3.9	Finding Distance of Two Points	37
3.10	Practices	38

4	**Linear Functions**	**45**
4.1	Determining Slopes	45
4.2	Formulating Linear Equations	46
4.3	Deriving Equations from Graphs	47
4.4	Understanding Slope-Intercept and Point-Slope Forms	48
4.5	Writing Point-Slope Equations from Graphs	49

4.6	Identifying x- and y-intercepts	50
4.7	Graphing Standard Form Equations	51
4.8	Understanding Horizontal and Vertical Lines	52
4.9	Graphing Horizontal or Vertical Lines	53
4.10	Graphing Point-Slope Form Equations	54
4.11	Understanding Parallel and Perpendicular Lines	56
4.12	Comparing Linear Function Graphs	57
4.13	Graphing Absolute Value Equations	58
4.14	Solving Two-Variable Word Problems	60
4.15	Practices	61

5 Inequalities and Systems of Equations 68

5.1	Solving One-Step Inequalities	68
5.2	Solving Multi-Step Inequalities	69
5.3	Working with Compound Inequalities	70
5.4	Graphing Solutions to Linear Inequalities	70
5.5	Writing Linear Inequalities from Graphs	72
5.6	Solving Advanced Linear Inequalities in Two Variables	73
5.7	Graphing Solutions to Advanced Linear Inequalities	74
5.8	Solving Absolute Value Inequalities	75
5.9	Understanding Systems of Equations	76
5.10	Determining the Number of Solutions to Linear Equations	77
5.11	Writing Systems of Equations from Graphs	78
5.12	Solving Systems of Equations Word Problems	80
5.13	Solving Word Problems Involving Linear Equations	81
5.14	Working with Systems of Linear Inequalities	81
5.15	Writing Word Problems for Two-Variable Inequalities	83

| 5.16 | Practices | 84 |

6 Polynomial . . . 98

6.1	Simplifying Polynomial Expressions	98
6.2	Adding and Subtracting Polynomial Expressions	99
6.3	Using Algebra Tiles to Add and Subtract Polynomials	100
6.4	Multiplying Monomial Expressions	101
6.5	Dividing Monomial Expressions	102
6.6	Multiplying a Polynomial by a Monomial	103
6.7	Using Area Models to Multiply Polynomials	104
6.8	Multiplying Binomial Expressions	105
6.9	Using Algebra Tiles to Multiply Binomials	105
6.10	Factoring Trinomial Expressions	106
6.11	Factoring Polynomial Expressions	107
6.12	Using Graphs to Factor Polynomials	108
6.13	Factoring Special Case Polynomial Expressions	110
6.14	Using Polynomials to Find Perimeter	110
6.15	Practices	111

7 Quadratic . . . 121

7.1	Solving Quadratic Equations	121
7.2	Graphing Quadratic Functions	122
7.3	Factoring to Solve Quadratic Equations	124
7.4	Understanding Transformations of Quadratic Functions	125
7.5	Completing Function Tables for Quadratic Functions	126
7.6	Determining Domain and Range of Quadratic Functions	127
7.7	Factoring Special Case Quadratics	128

7.8	Using Algebra Tiles to Factor Quadratics	129
7.9	Writing Quadratic Functions from Vertices and Points	130
7.10	Practices	131

8 Relations and Functions . . . 141

8.1	Understanding Function Notation and Evaluation	141
8.2	Completing a Function Table from an Equation	142
8.3	Determining Domain and Range of Relations	144
8.4	Performing Addition and Subtraction of Functions	144
8.5	Performing Multiplication and Division of Functions	145
8.6	Composing Functions	146
8.7	Evaluating Exponential Functions	147
8.8	Matching Exponential Functions with Graphs	147
8.9	Writing Exponential Functions from Word Problems	149
8.10	Understanding Function Inverses	150
8.11	Understanding Rate of Change and Slope	152
8.12	Practices	153

9 Radical Expressions . . . 162

9.1	Simplifying Radical Expressions	162
9.2	Performing Addition and Subtraction of Radical Expressions	163
9.3	Performing Multiplication of Radical Expressions	163
9.4	Rationalizing Radical Expressions	164
9.5	Solving Radical Equations	165
9.6	Determining Domain and Range of Radical Functions	166
9.7	Simplifying Radicals with Fractional Components	167
9.8	Practices	168

10 Rational Expressions 175

10.1 Simplifying Complex Fractions 175
10.2 Graphing Rational Functions 177
10.3 Performing Addition and Subtraction of Rational Expressions 179
10.4 Performing Multiplication of Rational Expressions 180
10.5 Performing Division of Rational Expressions 181
10.6 Evaluating Integers with Rational Exponents 182
10.7 Practices 183

11 Statistics and Probabilities 190

11.1 Calculating Mean, Median, Mode, and Range 190
11.2 Creating a Pie Graph 191
11.3 Analyzing Scatter Plots 193
11.4 Calculating and Interpreting Correlation Coefficients 195
11.5 Formulating the Equation of a Regression Line 196
11.6 Understanding Correlation and Causation 198
11.7 Understanding Permutations and Combinations 200
11.8 Solving Probability Problems 200
11.9 Practices 201

12 Praxis Algebra I (5162) Test Review and Strategies 210

12.1 The Praxis Algebra I Test Review 210
12.2 Praxis Algebra I (5162) Test-Taking Strategies 211

13 Practice Test 1 216

13.1 Practices 216
13.2 Answer Keys 235

13.3	Answers with Explanation	237
14	**Practice Test 2**	**255**
14.1	Practices	255
14.2	Answer Keys	275
14.3	Answers with Explanation	277

1. Fundamental and Building Blocks

1.1 Adding and Subtracting Integers

Integers include zero, natural numbers, and the negative of the natural numbers

$$\mathbb{Z} = \{\cdots, -3, -2, -1, 0, 1, 2, 3, \cdots\}.$$

To add and subtract integers, we must understand their absolute values and signs. When numbers have the same sign, you simply add their absolute values and keep the sign. However, when the signs differ, subtract the smaller absolute value from the larger, then keep the sign of the larger absolute value in the result.

🔔 Key Point

In adding or subtracting integers, first observe their signs. If they are alike, add the numbers and maintain the same sign. If different, subtract the smaller from the larger and take the sign of the larger. Also, remember that subtracting a negative is like adding a positive.

Example Add -4 and $+3$.

Solution: Because the signs are different, we need to subtract the absolute values: $4 - 3 = 1$. Since the larger absolute value (4) is negative, the result is -1.

Example Subtract $+7$ from -8.

Solution: Subtracting a positive number is the same as adding its negative, so we are actually adding -8 and -7. Since both numbers are negative, we add them to get -15.

1.2 Multiplying and Dividing Integers

Multiplication and division are very similar operations; in fact, you can think of division as a form of inverse multiplication.

> When multiplying or dividing, if both integers are positive or both are negative, the result is positive. But, if one integer is positive and the other negative, the result is negative.

These rules are true no matter the order of the numbers. This is due to the Commutative Property of Multiplication, which says that the product of two integers remains the same, regardless of their multiplication order.

 Multiply -3 and 5.

Solution: We are multiplying one negative integer and one positive integer, so the result is negative. Therefore, $-3 \times 5 = -15$.

 Divide -20 by -5.

Solution: We are dividing two negative integers, so the result is positive. Therefore, $-20 \div (-5) = 4$.

1.3 Order of Operations

One of the most basic principles of algebra is the order of operations, which dictates the sequence in which operations should be performed in a mathematical expression. It ensures we always get a unique result for an algebraic expression.

> Remember the order of operations with "PEMDAS": First, Parentheses; next, Exponents (like powers and square roots); then, Multiplication and Division (go left to right); and lastly, Addition and Subtraction (also left to right).

For operations on the same level, like several multiplications or divisions, or additions and subtractions, always move from left to right.

 Simplify the expression $7 + 3 \times 2 - 1$.

Solution: Following the order of operations (PEMDAS), we first do the multiplication: $7 + 6 - 1$.

Next, moving from left to right, we perform the addition and subtraction to get the final result: $13 - 1 = 12$.

So, $7 + 3 \times 2 - 1 = 12$.

 Simplify the expression $8 - 3 \times (2 + 3^2)$.

Solution: Following the order of operations, we first deal with the parentheses. Within the parentheses, we have addition and an exponent operation. According to PEMDAS, we perform the exponent first: $8 - 3 \times (2 + 9) = 8 - 3 \times 11$.

Next, we perform the multiplication: $8 - 33$.

Then the subtraction: $8 - 33 = -25$.

So, $8 - 3 \times (2 + 3^2) = -25$.

1.4 The Distributive Property

The Distributive Property is a fundamental principle in Algebra that describes how multiplication interacts with addition or subtraction.

It states that for any integers a, b, and c: $a \times (b + c) = a \times b + a \times c$. The property also applies to subtraction, $a \times (b - c) = a \times b - a \times c$. Think of it like this: If you have a number outside parentheses and a couple of numbers inside, you can multiply the outside number with each inside number, one by one.

Key Point

> The Distributive Property states that multiplying a number by a group of numbers added together is the same as doing each multiplication separately.

The Distributive Property allows us to remove parentheses by creating equivalent expressions, making simplification of algebraic expressions possible.

Example Simplify the expression $5 \times (4+3)$.

Solution: Applying the Distributive Property, we rewrite the expression as: $5 \times (4+3) = 5 \times 4 + 5 \times 3 = 20 + 15 = 35$.

 Example Simplify the expression $6 \times (2y-3)$.

Solution: Using the Distributive Property, we can expand the expression as: $6 \times (2y-3) = 6 \times 2y - 6 \times 3 = 12y - 18$.

1.5 Integers and Absolute Value

Absolute value of an integer is the distance of that number from zero on the number line, irrespective of the direction. It is always a non-negative value.

Key Point

The absolute value of an integer is represented as $|a|$, where a is the integer. If a is positive or zero, $|a| = a$, but if a is negative, $|a| = -a$.

This concept enables us to compare the magnitudes of numbers, regardless of their sign, and is fundamental for understanding inequalities, distance, and sequences in further studies.

 Example Calculate the absolute value of 5 and -5.

Solution: The absolute value of 5 is $|5| = 5$ because 5 is 5 units to the right of zero. The absolute value of -5 is $|-5| = 5$ because -5 is 5 units to the left of zero.

 Based on absolute value, which number is greater: -7 or 6?

Solution: The absolute value of -7 is $|-7| = 7$ and the absolute value of 6 is $|6| = 6$. Although -7 is less than 6, the magnitude (or absolute value) of -7 is greater than that of 6.

1.6 Proportional Ratios

A ratio expresses the relationship between two quantities. It is typically presented in the form a:b or $\frac{a}{b}$.

A proportion, on the other hand, states that two ratios are equivalent. It is an equation that equates two

1.7 Similarity and Ratios

ratios, such as $\frac{a}{b} = \frac{c}{d}$. This means that 'a' is to 'b' as 'c' is to 'd'.

Key Point

A ratio of 'a' to 'b' implies that for every 'a' units of the first quantity, there are 'b' units of the second quantity.

Key Point

In a proportion, if $\frac{a}{b} = \frac{c}{d}$, it implies that $ad = bc$ when we cross-multiply. This method, known as cross-multiplication, is a useful technique for solving problems with proportions.

Example The ratio of red balls to blue balls in a bag is 3:4. If there are 12 red balls, how many blue balls are in the bag?

Solution: The ratio of red to blue balls is 3:4 which means for every 3 red balls, there are 4 blue balls. Let's set up a proportion: $\frac{3}{4} = \frac{12}{x}$.

where x is the number of blue balls. $\frac{3 \times 4}{4 \times 4} = \frac{12}{x}$.

Solve for x, $x = 16$. Hence, there are 16 blue balls in the bag.

Example A map's scale is given as 1:50000, representing that 1 cm on the map corresponds to 50000 cm in real life. If a distance between two locations on the map measures 3.5 cm, what is the actual distance between these locations?

Solution: The map scale ratio is 1:50000. Set up a proportion involving the map distance and the required actual distance: $\frac{1}{50000} = \frac{3.5}{x}$, where x is the actual distance.

Using cross-multiplication gives $x = 3.5 \times 50000 = 175000$ cm.

Since it is common to express longer distances in kilometers, convert cm to km: 175000 cm = 1.75 km.

Hence, the actual distance is 1.75 km.

1.7 Similarity and Ratios

In algebra, the principle of 'similarity' refers to the concept of two figures having the same shape but differ in size. This principle is often connected with ratios. The ratio of corresponding sides in similar figures is called the 'scale factor'. If two figures are similar, the ratio between their corresponding lengths is consistent. This ratio, called the 'scale factor', can be understood as the measure by which the original figure is enlarged or reduced to form the similar figure.

🔔 Key Point

Similar figures have proportional corresponding sides and equal corresponding angles. The ratio of corresponding sides is the 'scale factor', which is consistent throughout.

🔔 Key Point

In similar figures, the ratio of perimeters is the same as the scale factor. However, the ratio of the areas is the square of the scale factor, while the ratio of volumes is the cube of the scale factor.

Example The sides of two similar triangles are in the ratio 2:3. If the perimeter of the smaller triangle is 30 units, what is the perimeter of the larger triangle?

Solution: In similar figures, the ratio of perimeters is the same as the scale factor. Here, it's 2:3. If the perimeter of the smaller triangle is 30 units, the perimeter of the larger one is: $\frac{3}{2} \times 30 = 45$ units.

Hence, the perimeter of the larger triangle is 45 units.

Example Two cubes have edge lengths in the ratio 1:4. If the surface area of the smaller cube is 24 square units, what is the surface area of the larger cube?

Solution: The ratio of the areas of similar figures is the square of the scale factor. So, the scale factor here is 1:4, and the square of the scale factor is $4^2 = 16$.

If the surface area of the smaller cube is 24 sq units, the surface area of the larger cube is: $16 \times 24 = 384$ sq units.

Hence, the surface area of the larger cube is 384 sq units.

1.8 Percent Problems

A percent is just a fraction whose denominator is 100. When we say "percent" we are really saying "per 100". One percent (1%) means 1 per 100.

1.9 Percent of Increase and Decrease

> **Key Point**
>
> A percent represents a fraction with a denominator of 100.

Percent problems usually have three components - the percent or rate ($r\%$), the base (b), and percentage the amount (p). You can use the formula $b \times \frac{r}{100} = p$ to solve percent problems. Depending on what you are asked to find, you will need to rearrange the formula and solve for the unknown.

> **Key Point**
>
> The formula to solve percent problems is $b \times \frac{r}{100} = p$. The base ($b$) is what you are taking the percent of. The rate (r) is the percent. The percentage amount (p) is the result of applying the percent to the base.

Example What is 20% of 50?

Solution: Here, it is clear that $r = 20$, $b = 50$, and you are asked to find p.
Using the formula $b \times \frac{r}{100} = p$, substitute the given values to find p: $50 \times \frac{20}{100} = p$.
Solving the equation gives $p = 10$. So, 20% of 50 is 10.

Example If 15% of a number is 30, what is the number?

Solution: Here, $r = 15$ is the rate, $p = 30$ is the percentage amount, and the base b is what you're solving for.
Rearranging the formula gives you $\frac{p}{\frac{r}{100}} = b$. Plug in the values known: $\frac{30}{\frac{15}{100}} = b$.
From this, you find that $b = 200$. So, the number is 200.

1.9 Percent of Increase and Decrease

In our daily life, we often come across phrases such as 'price has increased by a certain percent' or 'sales have dropped by a certain percent'. These are examples of percent increase and decrease, respectively.

Percent increase and decrease are measurements that express the relative growth or decline in value.

Chapter 1. Fundamental and Building Blocks

> **Key Point**
>
> Percent increase refers to how much the number has gone up as a percentage of the original. Percent decrease, on the contrary, refers to how much the number has reduced as a percentage of the original.

To calculate the percent increase:

> **Key Point**
>
> Percent Increase = $\frac{\text{Final Value} - \text{Initial Value}}{\text{Initial Value}} \times 100\%$.

To calculate the percent decrease:

> **Key Point**
>
> Percent Decrease = $\frac{\text{Initial Value} - \text{Final Value}}{\text{Initial Value}} \times 100\%$.

Example A watermelon costs $4 last month but now costs $5. By what percent did the price increase?

Solution: Here, the Initial Value is $4 and the Final Value is $5.

Using the percent increase formula:

Percent Increase = $\frac{5-4}{4} \times 100\%$.

Calculating the equation gives a 25% increase.

So, the price of the watermelon increased by 25%.

Example A shirt was originally priced at $50 but is now on sale for $40. By what percent did the price decrease?

Solution: Here, the Initial Value is $50 and the Final Value is $40.

Using the percent decrease formula:

Percent Decrease = $\frac{50-40}{50} \times 100\%$.

Calculating the equation gives a 20% decrease.

So, the price of the shirt decreased by 20%.

1.10 Discount, Tax, and Tip

Regularly, we encounter scenarios where we need to calculate discounts, tax, and tips. Maths skills are essential for understandable, accurate calculations in these financial contexts.

A **discount** refers to a decrease in the sale price of an item. The discount rate is usually specified as a percentage of the original price (also known as the mark price).

To calculate a discount: Discount = $\frac{\text{Discount Rate}}{100} \times$ Mark Price.

 Example If a pair of shoes originally cost $200 and are now 20% off, what is the discount?

Solution: Here, the Mark Price is $200 and the Discount Rate is 20%.

Using the discount formula:

Discount = $\frac{20}{100} \times 200 = \$40$.

So, the discount on the shoes is $40.

A **tax** is an additional amount of money charged on goods and services. It is also usually specified as a percentage of the price of the item.

To calculate tax: Tax = $\frac{\text{Tax Rate}}{100} \times$ Price.

 Example If a dining table cost $500 and a 7% sales tax is charged, how much is the tax?

Solution: Here, the Price is $500 and the Tax Rate is 7%.

Using the tax formula:

Tax = $\frac{7}{100} \times 500 = \$35$.

So, the tax on the dining table is $35.

A **tip** is an additional amount of money given to service staff. It is calculated as a percentage of the total bill.

> **Key Point**
>
> To calculate a tip: Tip = $\frac{\text{Tip Rate}}{100}$ × Total Bill.

Example If the total bill at a restaurant is $40 and you want to give a 15% tip, how much is the tip?

Solution: Here, the Total Bill is $40 and the Tip Rate is 15%.

Using the tip formula:

Tip = $\frac{15}{100} \times 40 = \$6$.

So, the tip on the $40 restaurant bill is $6.

1.11 Simple Interest

Simple Interest is a method of calculating interest on a loan or investment where the interest is calculated only on the initial amount (principal) that was deposited or borrowed.

> **Key Point**
>
> The formula for calculating simple interest is: $I = P \cdot R \cdot T$, where I is the interest, P the principal, R the rate of interest, and T the time in years.

Example Assume that you invested $2000 in a bank at an interest rate of 5% per annum. What will be the interest you earn after 3 years?

Solution: In this situation, the Principal $P = \$2000$, the Rate of interest $R = 0.05$, and the Time $T = 3$ years.

We calculate the interest I by using the formula for Simple interest:

$I = P \cdot R \cdot T = \$2000 \cdot 0.05 \cdot 3 = \$300$.

Therefore, the interest earned will be $300.

Example A loan of $1500 is borrowed at an interest rate of 6% per annum. What will be the interest to be paid after 2 years?

Solution: In this example, again we know the Principal $P = \$1500$, the Rate of interest $R = 0.06$, and time $T = 2$ years.

We calculate the interest I by using the formula for simple interest:

$I = P \cdot R \cdot T = \$1500 \cdot 0.06 \cdot 2 = \$180.$

Therefore, the interest to be paid after 2 years will be $180.

1.12 Approximating Irrational Numbers

Rational numbers are numbers that can be written as a fraction of two integers. *Irrational numbers* cannot be written as a fraction.

> **Key Point**
>
> Numbers that cannot be written as a fraction are called *irrational*. They have non-repeating, non-terminating decimal expansions.

One example is the square root of a number that is not a perfect square, such as $\sqrt{2}$ or $\sqrt{3}$. While you can never find an exact decimal or fractional value for these numbers, you can approximate their values for practical purposes.

Why Approximate? Approximating irrational numbers is crucial for various applications like engineering, science, and finance. We often need a number that is "good enough" for these purposes.

> **Key Point**
>
> The goal of approximation is to find a value close enough to the true value of an irrational number for practical usage.

 Approximate e to two decimal places.

Solution: The number e is an important irrational number that is approximately equal to 2.71828. To approximate e to two decimal places, we can round it to 2.72.

 Use rational numbers to approximate π for calculations.

Solution: π is another famous irrational number. For most practical purposes, you can approximate π as 3.14 or $\frac{22}{7}$.

1.13 Practices

Solve:

1) What is $-14 + (-7)$?

2) What is $13 - (-4)$?

3) What is $-3 + 7$?

4) What is $8 - 11$?

5) What is $-5 - (-2)$?

Solve the following problems:

6) Multiply -4 and 6.

7) Divide -42 by -7.

8) Multiply 10 and 3.

9) Divide 25 by -5.

10) Multiply -7 by -9.

Solve the following equations:

11) Solve $(6+4) \times 2 - 3^2$.

12) Solve $5 - 3 \times (1 + 2^3)$.

13) Solve $(8+2) - 3^2 \times 2$.

14) Solve $7 - 3 \times 2 + 4^2$.

15) Solve $(5^2 - 3) \times 2 - 1$.

1.13 Practices

Fill in the Blank:

16) The absolute value of an integer is its distance from _____ on the number line.

17) The absolute value of -9 is _____.

18) If an integer a is positive or zero, $|a| =$ _____.

19) Given an integer a, if $|a| = a$, a can be _____.

20) The number line representation of -5 is _____ units to the _____ of zero.

Solve:

21) If the ratio of cats to dogs in a pet shop is 5:7 and there are 35 dogs, how many cats are in the pet shop?

22) A map indicates that 2 cm corresponds to 100 km. If a distance between two locations on the map measures 10 cm, what is the actual distance between these locations in kilometers?

23) The ratio of purple marbles to green marbles in a bag is 4:3. If there are 32 green marbles, how many purple marbles are there?

24) An architect's blueprint of a house has a scale of 1:100, where 1 cm on the blueprint corresponds to 100 cm in actual size. If the length of the house on the blueprint is 8 cm, what is the actual length of the house in meters?

25) The ratio of weights of copper to zinc in a bronze alloy is 9:1. If there are 18 kg of zinc in the alloy, how much copper is present?

Solve:

26) What is 30% of 80?

27) If 25% of a number is 20, what is the number?

28) A car was originally priced at $14,000 but is now on sale for $11,200. What is the percent decrease?

29) The price of a house increased from $200,000 to $260,000. What was the percent increase?

30) You received a 60% discount on your $500 purchase. How much was the discount?

Fill in the Blank:

31) A dress that usually costs $100 is on sale for 15% off. The discount is _____.

32) A $150 teddy bear has a tax rate of 10%. The tax is _____.

33) If you want to leave a 20% tip on a $25 bill, the tip is _____.

34) A $800 laptop is on sale for 25% off. The discount is _____.

35) A $30 meal has a tax rate of 8%. The tax is _____.

True or False:

36) If you invest $1000 at an annual rate of 5% for 5 years, the interest earned would be $250.

37) A loan of $2000 taken at an annual rate of 4% for 3 years will accrue $240 in interest.

38) $5000 deposited at an annual rate of 6% for 4 years will earn $1200 in interest.

39) An investment of $3000 at an annual rate of 5% for 2 years will result in $400 of interest.

40) A $4000 loan at a rate of 7% per year for 4 years will accrue $840 in interest.

Answer Keys

1) −21
2) 17
3) 4
4) −3
5) −3
6) −24
7) 6
8) 30
9) −5
10) 63
11) 11
12) −22
13) −8
14) 17
15) 43
16) Zero
17) 9
18) a
19) Positive or Zero
20) 5, left
21) 25 cats
22) 500 km
23) 42 purple marbles
24) 8 meters
25) 162 kg.
26) 24
27) 80
28) 20%
29) 30%
30) $300
31) $15
32) $15
33) $5
34) $200
35) $2.40
36) True
37) True
38) True
39) False
40) False

Answers with Explanation

1) Since both numbers are negative, we add them to get -21.

2) Subtracting a negative is the same as adding its positive, so $13-(-4)$ equals $13+4=17$.

3) Since the numbers have different signs, we subtract their absolute values to get 4 and since $7>3$, the result is positive.

4) Subtract 11 from 8 to get -3.

5) Subtracting a negative is the same as adding its positive, so $-5-(-2)$ equals $-5+2=-3$.

6) We have a positive and a negative integer. Thus, the answer is negative. Therefore, $-4\times 6=-24$.

7) We are dividing two negative integers, so the result is positive. So, $-42\div -7=6$.

8) We have two positive integers. Thus, the answer is positive. Therefore, $10\times 3=30$.

9) We are dividing a positive and a negative integer, so the result is negative. So, $25\div -5=-5$.

10) We have two negative integers. Thus, the answer is positive. Therefore, $-7\times -9=63$.

11) First, simplify the parentheses: $10\times 2-3^2$. Then, the exponent operation: $10\times 2-9$. Next, perform the multiplication: $20-9$. Finally, the subtraction to get the final result: $20-9=11$.

12) First deal with the parentheses. In the parentheses, perform the exponent first: $5-3\times(1+8)$. Then simplification inside parentheses: $5-3\times 9$. After that, perform the multiplication: $5-27$. Finally, subtract: $5-27=-22$.

13) First, simplify the parentheses: $10-3^2\times 2$. Then, do the exponent operation: $10-9\times 2$. Next, perform the multiplication: $10-18$. Finally, subtract to yield the final result: $10-18=-8$.

14) First, do the exponent operation: $7-3\times 2+16$. Next, from left to right, perform the multiplication:

1.13 Practices

$7 - 6 + 16$. Then, moving from left to right, perform the operations: $1 + 16$. Finally, perform the addition to get the final result: $1 + 16 = 17$.

15) First, simplify the parentheses: $(25 - 3) \times 2 - 1$. Then do the subtraction in parentheses: $22 \times 2 - 1$. Next, perform the multiplication: $44 - 1$. Finally, subtract to yield the final result: $44 - 1 = 43$.

16) The absolute value of an integer is its distance from zero on the number line.

17) The absolute value of an integer a is represented as $|a|$, where $|a| = -a$ if a is negative. Thus, the absolute value of -9 is $|-9| = 9$.

18) For any integer a, if a is positive or zero, the absolute value of a is just a itself, i.e., $|a| = a$.

19) If the absolute value of an integer equals the integer itself ($|a| = a$), then the integer a can be positive or zero.

20) On a number line, -5 is represented 5 units to the left of zero. Negative integers are placed to the left of zero.

21) The ratio of cats to dogs is 5:7 implying for every 5 cats, there are 7 dogs. We set up the proportion $\frac{5}{7} = \frac{x}{35}$, where x is the number of cats. Cross-multiplying gives $5 \times 35 = 175$. Solving for x gives $x = \frac{175}{7} = 25$. Thus, there are 25 cats in the pet shop.

22) The map scale ratio is $\frac{2}{100}$. Let's set up a proportion involving the map distance and the required actual distance: $\frac{2}{100} = \frac{10}{x}$. Solving for x gives $x = \frac{100 \times 10}{2} = 500$. Thus, the actual distance between the two locations is 500 km.

23) The ratio of purple to green marbles is 4:3 implying for every 4 purple marbles, there are 3 green marbles. We set up the proportion $\frac{4}{3} = \frac{x}{32}$ where x is the number of purple marbles. Cross-multiplying gives $4 \times 32 = 128$. Solving for x gives $x = \frac{128}{3} \approx 42.67$. Since there cannot be a fraction of a marble, we truncate it, which gives us 42 purple marbles.

24) The blueprint scale ratio is $\frac{1}{100}$. Let's set up a proportion involving the blueprint length and the required actual length: $\frac{1}{100} = \frac{8}{x}$. Solving for x gives $x = \frac{100 \times 8}{1} = 800$. Thus, the actual length of the house is 800 cm which is equivalent to 8 meters.

Chapter 1. Fundamental and Building Blocks

25) The ratio of copper to zinc weights is 9:1 implying for every 9 kg of copper, there is 1 kg of zinc. We set up the proportion $\frac{9}{1} = \frac{x}{18}$, where x is the weight of copper. Cross-multiplying gives $9 \times 18 = 162$. Solving for x gives $x = 162$. Thus, there are 162 kg of copper in the alloy.

26) We use the formula $b \times \frac{r}{100} = p$, with $b = 80$ and $r = 30$. This gives us $80 \times \frac{30}{100} = 24$.

27) We rearrange the formula to find the base: $\frac{p}{r/100} = b$, with $p = 20$ and $r = 25$. This gives us some number $b = \frac{20}{25/100} = 80$.

28) The base (b) is $14,000 and the discount (p) is $2,800 ($14,000 - $11,200 = $2,800). By using the formula to calculate the rate, we get $r = \frac{p}{b} \times 100$, yielding $r = \frac{\$2,800}{\$14,000} \times 100 = 20\%$.

29) The base (b) is $200,000 and the increase (p) is $60,000 ($260,000 - $200,000). Then, using $r = \frac{p}{b} \times 100$, we find $r = \frac{\$60,000}{\$200,000} \times 100 = 30\%$.

30) We use the formula $b \times \frac{r}{100} = p$, with $b = \$500$ and $r = 60$ to find p, giving us $\$500 \times \frac{60}{100} = \$300$.

31) Discount = $\frac{\text{Discount Rate}}{100} \times$ Mark Price = $\frac{15}{100} \times 100 = \$15$

32) Tax = $\frac{\text{Tax Rate}}{100} \times$ Price = $\frac{10}{100} \times 150 = \$15$

33) Tip = $\frac{\text{Tip Rate}}{100} \times$ Total Bill = $\frac{20}{100} \times 25 = \$5$

34) Discount = $\frac{\text{Discount Rate}}{100} \times$ Mark Price = $\frac{25}{100} \times 800 = \$200$

35) Tax = $\frac{\text{Tax Rate}}{100} \times$ Price = $\frac{8}{100} \times 30 = \$2.40$

36) Using the simple interest formula, we find $I = P \cdot R \cdot T = \$1000 \times 0.05 \times 5 = \$250$.

37) Using the simple interest formula, we find $I = P \cdot R \cdot T = \$2000 \times 0.04 \times 3 = \$240$.

38) Using the simple interest formula, we find $I = P \cdot R \cdot T = \$5000 \times 0.06 \times 4 = \$1200$.

39) Using the simple interest formula, we find $I = P \cdot R \cdot T = \$3000 \times 0.05 \times 2 = \$300$, not $400.

40) Using the simple interest formula, we find $I = P \cdot R \cdot T = \$4000 \times 0.07 \times 4 = \$1120$, not $840.

2. Exponents and Variables

2.1 Mastering Exponent Multiplication

Exponentiation is just repeated multiplication. For example, 3^2 is just 3 times 3, which equals 9.

Key Point

When multiplying terms with the same base, you simply add the exponents together: $a^n \times a^m = a^{n+m}$. This is known as the *Product of Powers Property*.

Example What is the product of 5^3 and 5^4?

Solution: Since we are multiplying numbers with the same base, we add the exponents. Hence, $5^3 \times 5^4 = 5^{3+4} = 5^7$.

Example Simplify the product $(3 \times 5^4) \times (2 \times 5^3)$.

Solution: First, multiply the numbers 3 and 2 to get 6. Next, since we are multiplying two terms with base 5, we add their exponents 4 and 3 to get 7. Hence, the product is 6×5^7.

2.2 Exploring Powers of Products

Let's learn about multiplying numbers and raising them to a power. This involves a rule called the *Power of a Product Property*. This rule tells us that if we raise a product to a power, it is the same as raising each part of the product to that power.

Key Point

The Power of a Product Property: $(ab)^n = a^n b^n$.

 Example Simplify the expression $(5 \times 2)^3$.

Solution: From the Power of a Product property, we have $(5 \times 2)^3 = 5^3 \times 2^3$. Calculating these values, we get $5^3 = 125$ and $2^3 = 8$. So the simplified expression turns out to be $125 \times 8 = 1000$.

 Example Expand the expression $(3x)^4$.

Solution: By applying the Power of a Product property, we get $(3x)^4 = 3^4 \times x^4$. Simply calculate 3^4 to obtain 81, hence, $(3x)^4$ simplifies to $81x^4$.

2.3 Mastering Exponent Division

When we divide terms with the same base, we essentially subtract the exponent of the divisor from the exponent of the dividend.

Key Point

When dividing terms with the same base, subtract the exponents. Therefore, $b^m \div b^n = b^{m-n}$ if $b \neq 0$. This is known as the *Quotient of Powers Property*.

 Example Simplify the expression $\frac{7^6}{7^2}$.

Solution: As we are dividing numbers with the same base, we subtract the exponents. Hence, $\frac{7^6}{7^2} = 7^{6-2} = 7^4$.

 Example Simplify the expression $(10 \times 3^9) \div (2 \times 3^4)$.

Solution: First, divide the number 10 by 2 to get approximately 5. Next, since we are dividing two terms with base 3, we subtract the exponent of the divisor from the exponent of the dividend, which gives us $9 - 4 = 5$. Hence, the quotient is 5×3^5.

2.4 Understanding Zero and Negative Exponents

Regardless of the non-zero value of a we always have: $a^0 = 1$.

For any nonzero number a raised to the power of $-n$, where n is a positive integer, the expression simplifies to the reciprocal of a raised to the power of n. This means, a^{-n} equals $\frac{1}{a^n}$.

> **Key Point**
>
> The Zero Exponent Rule: For any non-zero numeral a, $a^0 = 1$.

> **Key Point**
>
> The Negative Exponent Rule: $a^{-n} = \frac{1}{a^n}$, where a is any non-zero number and n is a positive integer.

Example What is 3^0?

Solution: Using the Zero Exponent Rule, we can establish that any numeral raised to the power of zero equals 1. Therefore, $3^0 = 1$.

Example What is 10^{-2}?

Solution: Applying the Negative Exponent Rule, we know that $10^{-2} = \frac{1}{10^2}$. After calculating the denominator, we get $\frac{1}{10^2} = \frac{1}{100}$ or 0.01. Therefore, $10^{-2} = 0.01$.

2.5 Working with Negative Bases

When a base is negative, the result changes with even or odd exponents.

> **Key Point**
>
> $(-a)^n$ equals a^n if n is an even positive integer, and $-a^n$ if n is an odd positive integer.

An easy way to conceptualize this is to imagine yourself multiplying the negative base with itself n times. With an even exponent, the negatives will always cancel out, leaving us with a positive result. An odd exponent, however, will always have one negative left over, resulting in a negative conclusion.

 What is $(-2)^3$?

Solution: By applying the rule for when the base $-a$ is raised to an odd positive exponent value, the result is negative. Therefore: $(-2)^3 = -2^3 = -8$.

Remember, these rules only apply if the negative base is in parentheses. The placement of a negative sign greatly influences the result.

 What is -3^2 and how is it different from $(-3)^2$?

Solution: -3^2 becomes $-(3^2)$, which results in -9. In contrast, $(-3)^2$, where -3 is considered as a whole base, results in 9, following the rule for negative bases raised to even powers.

2.6 Introduction to Scientific Notation

Scientific notation is a method of writing numbers that are too big or too small to be conveniently written in decimal form. It is typically used in the fields of science and engineering, where dealing with immensely large or minutely small quantities is common.

The basic structure of scientific notation is $a \times 10^n$, where $1 \leq |a| < 10$ and n is an integer. Here, a is called the mantissa and n is the exponent.

🔔 **Key Point**

> Scientific notation involves expressing a number as a product of a number between 1 and 10 and an appropriate power of 10.

 Express 3000 and 0.0056 in scientific notation.

Solution: For 4000, shifting the decimal point 3 places to the left gives 4. Number of places shifted becomes our exponent, which is 3. Therefore, 4000 in scientific notation is 4×10^3.

For 0.0056, shifting the decimal point 3 places to the right will give 5.6. As we're dealing with a number smaller than 1, the exponent will be the negative of the number of places shifted, giving us -3. Hence, 0.0056 in scientific notation is 5.6×10^{-3}.

It is crucial to remember that the sign of the exponent indicates the direction of the original number's

2.7 Addition and Subtraction in Scientific Notation

decimal shift, while its magnitude represents the number of places the decimal must move.

 Example What are the mantissa and exponent for the number 7.35 in scientific notation?

Solution: In this case, the number 7.35 is already between 1 and 10 and thus forms our mantissa directly. As we didn't need to shift the decimal point, the exponent corresponding to the power of 10 is 0. Therefore, 7.35 in scientific notation is 7.35×10^0.

2.7 Addition and Subtraction in Scientific Notation

To add or subtract two numbers that are presented in scientific notation, the exponents of each must be the same.

> To add or subtract numbers in scientific notation, ensure the exponents for each number are the same.

If the numbers to be added or subtracted have different exponents, adjust one or both numbers so that the exponents match. Once the exponents are equal, you can directly add or subtract the mantissas.

 **Example** Add 2×10^5 and 5×10^4.

Solution: First, align the exponents. Rewrite 5×10^4 as 0.5×10^5. Now the exponents are the same, we can add the mantissas: $2 \times 10^5 + 0.5 \times 10^5 = (2 + 0.5) \times 10^5 = 2.5 \times 10^5$.

Just like with addition, it is important to have the same exponents when you are subtracting numbers in scientific notation.

 Example Subtract 4×10^6 from 9×10^7.

Solution: Align the exponents first. Convert 4×10^6 to 0.4×10^7. Now, subtract the mantissas: $9 \times 10^7 - 0.4 \times 10^7 = 8.6 \times 10^7$.

2.8 Multiplication and Division in Scientific Notation

When multiplying numbers expressed in scientific notation, we multiply the mantissas (the numbers before "times 10") and add the exponents. In contrast, for division, we divide the mantissas and subtract the exponents.

Key Point

To multiply numbers in scientific notation, multiply the mantissas and add the exponents. For division, divide the mantissas and subtract the exponents.

Example Multiply 2×10^5 and 5×10^3.

Solution: Multiply the mantissas: $2 \times 5 = 10$. Add the exponents: $5 + 3 = 8$. So, the answer is 10×10^8, which we adjust to 1×10^9 for proper scientific notation.

Example Divide 1.2×10^{-2} by 3×10^5.

Solution: Divide the mantissas: $1.2 \div 3 = 0.4$. Subtract the exponents: $-2 - 5 = -7$. The answer is 0.4×10^{-7}.

2.9 Practices

Solve:

1) Simplify $3^5 \times 3^7$.

2) Solve for x in the equation $2^4 \times 2^x = 2^9$.

3) Simplify $(1.5 \times 10^4) \times (3.0 \times 10^5)$.

4) If $a = 2^7$, then what is the value of $a \times 2^3$?

5) Simplify $7^4 \times 7^2$.

Simplify Each Expression:

6) Simplify $(4 \times 7)^2$.

7) Simplify $(3 \times 5)^4$.

8) Simplify $(6 \times 2)^3$.

2.9 Practices

9) Simplify $(9 \times 3)^2$.

10) Simplify $(10 \times 8)^3$.

Fill in the Blank:

11) $5^8 \div 5^\square = 5^5$

12) $\frac{2^{10}}{2^\square} = 2^4$

13) $\frac{7^\square}{7^2} = 7^3$

14) $\frac{11^7}{11^\square} = 11^5$

15) $\frac{3^{12}}{3^\square} = 3^7$

Solve:

16) Solve 4^0.

17) Solve 6^{-2}.

18) Solve 8^0.

19) Solve 7^{-1}.

20) Solve $2^5 \div 2^9$.

Fill in the Blanks:

21) If n is even, and a is any real number, then $(-a)^n = $ _____.

22) If n is odd, and a is any real number, then $(-a)^n = $ _____.

23) Calculate the value of $(-7)^2$.

24) Calculate the value of $(-5)^3$.

Chapter 2. Exponents and Variables

25) Calculate the value of -4^2 and explain how it is different from $(-4)^2$.

Solve:

26) Write 25000 in scientific notation.

27) Write 6.72×10^{-7} in standard notation.

28) Write -0.056 in scientific notation.

29) Write 2.874 in scientific notation.

30) Write 607×10^3 in correct scientific notation.

Fill in the Blank:

31) Adjust the exponents to add in scientific notation: $1.2 \times 10^3 + 5 \times 10^2 = $ _____ $\times 10^3$.

32) Add these in scientific notation after adjusting the exponents: $3.6 \times 10^2 + 4 \times 10^1 = $ _____ $\times 10^2$.

33) Correctly add after matching exponents: $11 \times 10^5 + 3 \times 10^6 = $ _____ $\times 10^6$.

34) Calculate this after adjusting exponents: $2.8 \times 10^7 + 7 \times 10^6 = $ _____ $\times 10^7$.

35) Solve after adjusting exponents: $9.5 \times 10^4 + 2 \times 10^5 = $ _____ $\times 10^5$.

Solve:

36) Multiply 3.2×10^6 and 4.2×10^7.

37) Multiply 6.7×10^3 and 2.3×10^5.

38) Multiply 8.5×10^2 and 5.6×10^3.

39) Multiply 7.7×10^4 and 3.3×10^6.

40) Multiply 1.2×10^8 and 5.5×10^9.

2.9 Practices

Answer Keys

1) 3^{12}
2) $x = 5$
3) 4.5×10^9
4) 2^{10}
5) 7^6
6) 784
7) 50625
8) 1728
9) 729
10) 512000
11) 3
12) 6
13) 5
14) 2
15) 5
16) 1
17) $\frac{1}{36}$
18) 1
19) $\frac{1}{7}$
20) 0.0625
21) a^n
22) $-a^n$
23) 49
24) -125
25) -16 and 16
26) 2.5×10^4
27) 0.000000672
28) -5.6×10^{-2}
29) 2.874×10^0
30) 6.07×10^5
31) 1.7×10^3
32) 4×10^2
33) 4.1×10^6
34) 3.5×10^7
35) 2.95×10^5
36) 1.344×10^{14}
37) 1.541×10^9
38) 4.76×10^6
39) 2.541×10^{11}
40) 6.6×10^{17}

Answers with Explanation

1) Since we are multiplying numbers with the same base, we add the exponents. Hence, $3^5 \times 3^7 = 3^{5+7} = 3^{12}$.

2) By the property of multiplying exponents, $2^4 \times 2^x = 2^{4+x}$. Hence, $4+x=9$. Solving this equation gives $x=5$.

3) First, multiply the numbers 1.5 and 3.0 to get 4.5. Next, since we are multiplying two terms with base 10, we add their exponents 4 and 5 to get 9. Hence, the product is 4.5×10^9.

4) Substituting $a=2^7$, the expression $a \times 2^3$ equals $2^7 \times 2^3$. By adding the exponents, we get 2^{10}.

5) Since we are multiplying numbers with the same base, we add the exponents. Hence, $7^4 \times 7^2 = 7^{4+2} = 7^6$.

6) $(4 \times 7)^2 = 4^2 \times 7^2 = 16 \times 49 = 784$.

7) $(3 \times 5)^4 = 3^4 \times 5^4 = 81 \times 625 = 50625$.

8) $(6 \times 2)^3 = 6^3 \times 2^3 = 216 \times 8 = 1728$.

9) $(9 \times 3)^2 = 9^2 \times 3^2 = 81 \times 9 = 729$.

10) $(10 \times 8)^3 = 10^3 \times 8^3 = 1000 \times 512 = 512000$.

11) To make the exponents subtract to 5, the blank should be filled with 3. Hence, $5^8 \div 5^3 = 5^{8-3} = 5^5$.

12) The exponent in the denominator must be 6, for the terms to divide out to 2^4. This is because $2^{10} \div 2^6 = 2^{10-6} = 2^4$.

13) By filling in 5 in the blank, we complete the expression, as $7^5 \div 7^2 = 7^{5-2} = 7^3$.

14) By the Quotient of Powers Property, we subtract the exponents, which gives us $11^7 \div 11^2 = 11^{7-2} = 11^5$.

15) The missing exponent is 5, as it gives $3^{12} \div 3^5 = 3^{12-5} = 3^7$.

2.9 Practices

16) As per the Zero Exponent Rule, any non-zero numeral raised to the power of zero equals 1.

17) Applying the Negative Exponent Rule, we find that $6^{-2} = \frac{1}{6^2} = \frac{1}{36}$.

18) Based on the Zero Exponent Rule, any non-zero numeral raised to the power of zero is equal to 1.

19) Applying the Negative Exponent Rule, we find that $7^{-1} = \frac{1}{7}$.

20) $2^5 \div 2^9 = 2^{-4} = \frac{1}{2^4} = \frac{1}{16} = 0.0625$.

21) According to the given rule, if the base $-a$ is raised to an even positive exponent, the result is positive a^n.

22) According to the given rule, if the base $-a$ is raised to an odd positive exponent, the result is $-a^n$.

23) According to the rule mentioned, when -7 is raised to an even positive exponent value, the result is $7^2 = 49$.

24) According to the rule mentioned, when -5 is raised to an odd positive exponent value, the result is $-5^3 = -125$.

25) -4^2 becomes $-(4^2)$ which is equal to -16, while $(-4)^2$, where -4 is considered as a single base, it results in 16, following the stated rule with even exponents.

26) For 25000, moving the decimal four places to the left gives us 2.5. The number of places we moved the decimal becomes the exponent, so we obtain 2.5×10^4.

27) For 6.72×10^{-7}, moving the decimal 7 places to the left (because exponent is negative) gives us 0.000000672.

28) For -0.056, moving the decimal two places to the right gives us -5.6. The number of places we moved the decimal becomes the negative exponent, so we obtain -5.6×10^{-2}.

29) For 2.874, no movement of the decimal point is required as it is already between 1 and 10. So, the exponent, n, is 0.

30) Moving the decimal point two places left in 607 gives us 6.07. Adding 2 to the existing exponent 3 (because we moved decimal two places to the left) gives us 5, so we obtain 6.07×10^5.

31) We adjust 5×10^2 to 0.5×10^3, then add the mantissas: $1.2 + 0.5 = 1.7$.

32) We adjust 4×10^1 to 0.4×10^2, then add the mantissas: $3.6 + 0.4 = 4$.

33) We adjust 11×10^5 to 1.1×10^6, then add the mantissas: $1.1 + 3 = 4.1$.

34) We adjust 7×10^6 to 0.7×10^7, then add the mantissas: $2.8 + 0.7 = 3.5$.

35) We adjust 9.5×10^4 to 0.95×10^5, then add the mantissas: $0.95 + 2 = 2.95$.

36) First, we multiply the mantissas (3.2 and 4.2) to get 13.44 and add the exponents (6 and 7) to get 13. The number 13.44 is not properly in scientific notation, so we adjust it by moving the decimal place one position to the left, and increase the exponent by 1 to account for this. Thus, 13.44×10^{13} becomes 1.344×10^{14}.

37) We start by multiplying the mantissas (6.7 and 2.3) to get 15.41, and adding the exponents (3 and 5) to get 8. Since the mantissa is not in proper scientific notation, we move the decimal one place to the left, getting 1.541, and increment the exponent by 1, giving us 1.541×10^9.

38) We multiply the mantissas (8.5 and 5.6) to get 47.6 and add the exponents (2 and 3) to get 5. The mantissa should be adjusted by moving one decimal place to the left, and the exponent also increased by 1. So, the answer is 4.76×10^6.

39) The mantissas (7.7 and 3.3) are multiplied to get 25.41, and the exponents (4 and 6) are added to get 10. Adjust the mantissa to 2.541 and increment the exponent to 11. So, the answer is 2.541×10^{11}.

40) By multiplying the mantissas (1.2 and 5.5), we get 6.6. By adding the exponents (8 and 9), we get 17. The answer in the appropriate scientific notation is 6.6×10^{17}.

3. Expressions and Equations

3.1 Translate a Phrase into an Algebraic Statement

An algebraic expression is an expression that includes constant numbers, variables and algebraic operations such as addition, subtraction, multiplication, division and exponentiation. For example $2x - y^2 + 1$.

Translating verbal phrases into algebraic statements allows you to model real-world problems mathematically, enabling solving them systematically.

Key Point

The first step in translating a phrase into an algebraic statement is identifying the variables. Next, recognize the operations (addition, subtraction, multiplication, or division) represented by phrases or words.

Often, the phrase 'the sum of' represents addition, 'less' represents subtraction, 'product of' represents multiplication, and 'quotient of' refers to division. Classifying expressions as equals is generally indicated by the term 'is' or 'are'.

Example

Translate the phrase "The sum of a number and six is twenty" into an algebraic statement.

Solution: Firstly, we use n as the variable. 'The sum of' translates to addition, 'and' usually signals the second term involved in the operation, and 'is' translates to equals. So, the phrase converts to the algebraic statement: $n + 6 = 20$.

Remember, sequential word positioning may not always represent the order of operations in a phrase.

Understanding operation-indicating words and proper application of mathematics are key.

Example Translate "Two increased by the product of three and a number (m) is eight" into an algebraic statement.

Solution: Here, 'm' works as the variable, 'Two increased by' translates to addition of two, 'the product of three and a number' translates to multiplication of three and the variable, and 'is' translates to equals. Thus, the phrase translates to $2 + 3m = 8$.

3.2 Simplifying Variable Expressions

When simplifying expressions, you must adhere to the Order of Operations: Firstly Parenthesis, Exponents, next Multiplication and Division (from left to right), and finally Addition and Subtraction (from left to right). This is often remembered using the abbreviation PEMDAS.

Key Point

Always follow the order of operations, commonly expressed as PEMDAS: Parenthesis, Exponents, Multiplication and Division, Addition and Subtraction.

Example Simplify $7x + 3x - 5$.

Solution: Combine the like terms: $7x$ and $3x$. This leaves us with: $7x + 3x - 5 = 10x - 5$. The simplified expression is $10x - 5$.

Example Simplify $4(x+5) + 3(2x-1)$.

Solution: Use the distributive property. Distribute 4 across $x + 5$ and 3 across $2x - 1$. This leaves us with: $4(x+5) + 3(2x-1) = 4x + 20 + 6x - 3$. Then combine like terms: $4x$ and $6x$, and 20 and -3: $4x + 20 + 6x - 3 = 10x + 17$. Thus, the simplified expression is $10x + 17$.

3.3 Evaluating Single Variable Expressions

When we speak of evaluating an expression, we refer to finding the numerical value of the expression by replacing the variable(s) in the expression with certain values.

Let's consider a single variable expression $4x+7$. If we want to evaluate this expression for $x = 3$, we replace the x in the equation with 3, and then follow the simplification rules as before.

> **Key Point**
>
> To evaluate a single variable expression, substitute the given value of the variable into the expression and simplify.

Example Evaluate $2x - 5$ for $x = 6$.

Solution: To evaluate this expression for $x = 6$, we replace x with 6. This gives us: $2x - 5 = 2(6) - 5 = 12 - 5 = 7$.

So, $2x - 5$ evaluates to 7 when $x = 6$.

Example Evaluate $4x^2 - 3x + 2$ for $x = 2$.

Solution: Replace x with 2, and calculate: $4x^2 - 3x + 2 = 4(2)^2 - 3(2) + 2 = 4(4) - 6 + 2 = 16 - 6 + 2 = 12$.

So, $4x^2 - 3x + 2$ evaluates to 12 when $x = 2$.

3.4 Evaluating Two Variable Expressions

We can evaluate two-variable expressions by replacing the variables with given values, then simplify the expression according to the operations' rules.

> **Key Point**
>
> To evaluate a two-variable expression, replace each variable with its given value and simplify. The order of substitution does not affect the result.

Example Evaluate $3x + 2y$ for $x = 4$ and $y = 2$.

Solution: To evaluate the expression for the given values, we substitute x and y with 4 and 2 respectively. $3x + 2y = 3(4) + 2(2) = 12 + 4 = 16$.

So, $3x + 2y$ evaluates to 16 when $x = 4$ and $y = 2$.

 **Example** Evaluate $x^2 - y^2$ for $x = 3$ and $y = 2$.

Solution: We substitute x for 3 and y for 2, and then perform the operations: $x^2 - y^2 = (3)^2 - (2)^2 = 9 - 4 = 5$.

So, $x^2 - y^2$ evaluates to 5 when $x = 3$ and $y = 2$.

3.5 Solving One-Step Equations

A one-step equation is an algebraic equation that can be solved in just one step. The goal is to work out what the variable represents.

One way to solve one-step equations is by addition or subtraction. Remember, what you do to one side of the equation, you must also do to the other side. This ensures that they both remain equal to each other.

 Key Point

For one-step addition or subtraction equations, to get the variable on one side and alone, perform the inverse operation on both sides of the equation.

 **Example** Solve the equation $x + 7 = 12$.

Solution: To isolate x, subtract 7 from both sides of the equation: $x + 7 - 7 = 12 - 7 \Rightarrow x = 5$.

Therefore, the solution to the equation $x + 7 = 12$ is $x = 5$.

Another way to solve one-step equations is by multiplication or division. If the equation involves multiplication, use division to isolate the variable. If the equation involves division, use multiplication.

 Key Point

For one-step multiplication or division equations, to get the variable on one side and alone, perform the inverse operation on both sides of the equation.

 **Example** Solve the equation $5x = 20$.

Solution: To isolate x, divide both sides of the equation by 5: $\frac{5x}{5} = \frac{20}{5} \Rightarrow x = 4$.

Therefore, the solution to the equation $5x = 20$ is $x = 4$.

3.6 Solving Multi-Step Equations

Multi-step equations are equations that require more than one operation, like addition, subtraction, multiplication, or division, to solve. The aim is to get the variable on its own on one side, but you might have to perform several steps and use your understanding of the order of operations to find the solution.

> **Key Point**
>
> The key to solving multi-step equations is performing one operation at a time, and always keeping the equation balanced by doing the same operation to both sides.

 Example Solve the equation $2x + 3 = 9$.

Solution: To begin to isolate x, start by subtracting 3 from both sides of the equation: $2x + 3 - 3 = 9 - 3 \Rightarrow 2x = 6$.

Next, to solve for x, divide both sides of the equation by 2: $\frac{2x}{2} = \frac{6}{2} \Rightarrow x = 3$.

Therefore, the solution to the equation $2x + 3 = 9$ is $x = 3$.

Example Solve the equation $3x - 4 = 14$.

Solution: First, add 4 to both sides of the equation to isolate the term with x: $3x - 4 + 4 = 14 + 4 \Rightarrow 3x = 18$.

Then, divide both sides by 3 to get x alone: $\frac{3x}{3} = \frac{18}{3} \Rightarrow x = 6$.

Therefore, the solution to the equation $3x - 4 = 14$ is $x = 6$.

3.7 Rearranging Equations with Multiple Variables

In algebra, we often encounter equations with multiple variables. To solve these equations for a particular variable, we need to rearrange them. This process involves using the properties of equality, along with addition, subtraction, multiplication, and division, to isolate the desired variable.

Key Point

The principle of balance is critical when rearranging equations with multiple variables. Any operation performed on one side of the equation must also be performed on the other to maintain equality.

 **Example** Rearrange the equation $5x - 3y = 2y + 7$ for the variable x.

Solution: To solve for x, we first want to gather all terms containing x on one side of the equation and all other terms on the other. We can start by adding $3y$ to both sides of the equation: $5x - 3y + 3y = 2y + 7 + 3y \Rightarrow 5x = 5y + 7$.

Next, to get x alone, divide the entire equation by 5: $\frac{5x}{5} = \frac{5y+7}{5} \Rightarrow x = y + \frac{7}{5}$. Therefore, the equation $5x - 3y = 2y + 7$ becomes $x = y + \frac{7}{5}$ when solved for x.

 **Example** Rearrange the equation $4a + 2b = 3a - b$ for the variable a.

Solution: Start with arranging all terms containing a on one side and all terms containing b on the other: $4a - 3a = -2b - b \Rightarrow a = -3b$.

Therefore, the equation $4a + 2b = 3a - b$ becomes $a = -3b$ when solved for a.

3.8 Finding Midpoint

The midpoint of a line is like the average position between its two ends.

To find a line segment's midpoint, we use a special formula. It averages the x and y coordinates of the segment's endpoints.

 Key Point

To find the midpoint, M, of a line with endpoints $A(x_1, y_1)$ and $B(x_2, y_2)$, the formula is:

$$M = \left(\frac{x_1 + x_2}{2}, \frac{y_1 + y_2}{2} \right).$$

 **Example** Find the midpoint of the line segment with the given endpoints $(2, -4)$, $(6, 8)$.

Solution: To find the midpoint, we use the midpoint formula substituting $x_1 = 2$, $y_1 = -4$, $x_2 = 6$, $y_2 = 8$:

$$M = \left(\frac{2+6}{2}, \frac{-4+8}{2} \right) = \left(\frac{8}{2}, \frac{4}{2} \right) = (4, 2).$$

Hence, the midpoint of the line segment with endpoints $(2, -4)$ and $(6, 8)$ is $(4, 2)$.

3.9 Finding Distance of Two Points

The distance between two points can be calculated using a formula.

To find the distance between $A(x_1, y_1)$ and $B(x_2, y_2)$, the formula is:

$$d = \sqrt{(x_2 - x_1)^2 + (y_2 - y_1)^2}.$$

Key Point

The order of points does not change the result. The distance from point A to point B is the same as from point B to point A.

Example Find the distance between $(4, 2)$ and $(-5, -10)$ on the coordinate plane.

Solution: Use the distance formula: $d = \sqrt{(x_2 - x_1)^2 + (y_2 - y_1)^2}$, where $(x_1, y_1) = (4, 2)$ and $(x_2, y_2) = (-5, -10)$.

Substituting these values into the formula gives us:

$$d = \sqrt{(-5-4)^2 + (-10-2)^2} = \sqrt{(-9)^2 + (-12)^2} = \sqrt{81 + 144} = \sqrt{225} = 15.$$

So, the distance between point A and point B is $d = 15$.

3.10 Practices

Translate the given phrase into an algebraic statement:

1) The sum of ten and a number a is thirty.

2) Two times a number b decreased by five is twelve.

3) The difference between seven times a number x and three is twenty.

4) The quotient of a number y and six is eight.

5) Three less than five times a number z is twenty.

Simplify Each Expression:

6) $8x - 3x + 9$

7) $5(2y - 1) + 3(y + 4)$

8) $7a + 3(2a - 5)$

9) $2(3k + 7) + 4(k - 1)$

10) $9 - 4b + 3(2b - 3)$

Find the value of y:

11) Solve $y = 3x + 4$ for $x = 3$.

12) Solve $y = 7x^2 - 4x + 6$ for $x = 2$.

13) Solve $y = 3x^3 - x^2 + 5x - 7$ for $x = 1$.

14) Solve $y = 2x - 9$ for $x = 3$.

15) Solve $y = 6x^2 + 3x - 10$ for $x = 2$.

3.10 Practices

Evaluate:

16) Evaluate $2x - y$ for $x = 5$ and $y = 3$.

17) Evaluate $4x + 3y$ when $x = -1$ and $y = 2$.

18) Evaluate $3x - 5y$ when $x = -2$ and $y = 1$.

19) Evaluate $5x + 10y$ if $x = 2$ and $y = 0$.

20) Evaluate $x^2 - y$ when $x = 3$ and $y = 5$.

Solve:

21) Solve the equation $y - 6 = 9$.

22) Solve the equation $8p = 64$.

23) Solve the equation $x + 4 = 0$.

24) Solve the equation $3z = 15$.

25) Solve the equation $\frac{m}{9} = 7$.

Solve the following multi-step equations:

26) Solve the equation $4x + 7 = 19$.

27) Solve the equation $5x - 12 = 18$.

28) Solve the equation $7x + 9 = 23$.

29) Solve the equation $6x - 10 = 8$.

30) Solve the equation $8x + 5 = 21$.

Solve the following equations for the specified variable:

31) Rearrange the equation $7x - 4y = 2x + 3$ for the variable x.

32) Rearrange the equation $3m - n = 4m + 5n$ for the variable m.

33) Rearrange the equation $5a - 2b = 7b + 3$ for the variable a.

34) Rearrange the equation $4p - 3q = q + 2$ for the variable p.

35) Rearrange the equation $4s + 3t = 2s - t$ for the variable s.

Answer Keys

1) $a + 10 = 30$
2) $2b - 5 = 12$
3) $7x - 3 = 20$
4) $y \div 6 = 8$
5) $5z - 3 = 20$
6) $5x + 9$
7) $13y + 7$
8) $13a - 15$
9) $10k + 10$
10) $2b$
11) 13
12) 26
13) 0
14) -3
15) 20
16) 7
17) 2
18) -11
19) 10
20) 4
21) $y = 15$
22) $p = 8$
23) $x = -4$
24) $z = 5$
25) $m = 63$
26) $x = 3$
27) $x = 6$
28) $x = 2$
29) $x = 3$
30) $x = 2$
31) $x = \frac{4y}{5} + \frac{3}{5}$
32) $m = -6n$
33) $a = \frac{9b+3}{5}$
34) $p = q + \frac{1}{2}$
35) $s = -2t$

Answers with Explanation

1) 'The sum of' translates to addition, hence $a+10$. 'is' translates to equals hence the statement becomes $a+10=30$.

2) 'Two times a number' translates to $2b$, 'decreased by' translates to subtraction hence $2b-5$. 'is' translates to equals hence the statement becomes $2b-5=12$.

3) 'Seven times a number' translates to $7x$, 'difference' translates to subtraction hence $7x-3$. 'is' translates to equals hence the statement becomes $7x-3=20$.

4) 'The quotient of' translates to division, hence $y \div 6$. 'is' translates to equals hence the statement becomes $y \div 6 = 8$.

5) 'Five times a number' translates to $5z$, 'less than' translates to subtraction hence $5z-3$. 'is' translates to equals hence the statement becomes $5z-3=20$.

6) Combine the like terms: $8x$ and $-3x$ which results in $5x$. The simplified expression is $5x+9$.

7) Apply the distributive property first. Then combine like terms to get the simplified expression $13y+7$.

8) Use the distributive property and combine like terms to get the simplified expression $13a-15$.

9) Apply the distributive property first, then combine the like terms to get the simplified expression $10k+10$.

10) Use the distributive property. Then combine like terms to get the simplified expression $2b$.

11) Substitute $x=3$ into $y=3x+4$. This gives $y=3(3)+4=9+4=13$.

12) Substitute $x=2$ into $y=7x^2-4x+6$. This gives $y=7(2)^2-4(2)+6=28-8+6=26$.

13) Substitute $x=1$ into $y=3x^3-x^2+5x-7$. This gives $y=3(1)^3-1+5-7=0$.

14) Substitute $x=3$ into $y=2x-9$. This gives $y=2(3)-9=-3$.

3.10 Practices

15) Substitute $x = 2$ into $y = 6x^2 + 3x - 10$. This gives $y = 6(2)^2 + 3(2) - 10 = 24 + 6 - 10 = 20$.

16) $2x - y = 2(5) - 3 = 10 - 3 = 7$.

17) $4x + 3y = 4(-1) + 3(2) = -4 + 6 = 2$.

18) $3x - 5y = 3(-2) - 5(1) = -6 - 5 = -11$.

19) $5x + 10y = 5(2) + 10(0) = 10 + 0 = 10$.

20) $x^2 - y = (3)^2 - 5 = 9 - 5 = 4$.

21) Add 6 on both sides of the equation to isolate y: $y - 6 + 6 = 9 + 6 \Rightarrow y = 15$. So, the solution is $y = 15$.

22) Divide both sides by 8 to isolate p: $\frac{8p}{8} = \frac{64}{8} \Rightarrow p = 8$. So, the solution is $p = 8$.

23) Subtract 4 from both sides to isolate x: $x + 4 - 4 = 0 - 4 \Rightarrow x = -4$. So, the solution is $x = -4$.

24) Divide both sides by 3 to isolate z: $\frac{3z}{3} = \frac{15}{3} \Rightarrow z = 5$. So, the solution is $z = 5$.

25) Multiply both sides by 9 to isolate m: $\frac{m}{9} \times 9 = 7 \times 9 \Rightarrow m = 63$. So, the solution is $m = 63$.

26) Subtract 7 from both sides of the equation, we have $4x = 12$. Then divide both sides by 4 to get $x = 3$.

27) Add 12 to both sides of the equation, we have $5x = 30$. Then divide both sides by 5 to get $x = 6$.

28) Subtract 9 from both sides of the equation, we have $7x = 14$. Then divide both sides by 7 to get $x = 2$.

29) Add 10 to both sides of the equation, we have $6x = 18$. Then divide both sides by 6 to get $x = 3$.

30) Subtract 5 from both sides of the equation, we have $8x = 16$. Then divide both sides by 8 to get $x = 2$.

31) First, subtract $2x$ from both sides to put all terms containing x on the left side of the equation: $7x - 2x = 4y + 3 \Rightarrow 5x = 4y + 3$. Thereafter, divide both sides by 5 to isolate x: $x = \frac{4y}{5} + \frac{3}{5}$.

32) Start by bringing terms containing m to one side and n on the other: $3m - 4m = n + 5n \Rightarrow -m = 6n$. Now, multiply both sides by -1 to get $m = -6n$.

33) First, add $2b$ to both sides to assemble all terms containing b on the right side: $5a = 9b + 3$. Thereafter, divide both sides by 5 to get a: $a = \frac{9b+3}{5}$.

34) First, add $3q$ to both sides of the equation: $4p = 4q + 2$. Then, divide both sides by 4 to isolate p: $p = q + \frac{1}{2}$.

35) Begin with subtracting $2s$ and adding $3t$ to both sides: $4s - 2s = -t - 3t \Rightarrow 2s = -4t$. Finally, divide both sides by 2 to isolate s: $s = -\frac{4t}{2} = -2t$.

4. Linear Functions

4.1 Determining Slopes

The slope of a line is a measure of the steepness or inclination of the line. In simple terms, it tells us how slanted a line is. The slope of a line is represented by the letter m.

> **Key Point**
>
> The slope of a line passing through two points $A(x_1, y_1)$ and $B(x_2, y_2)$ is given by the formula: $m = \frac{y_2 - y_1}{x_2 - x_1}$.

The numerator $y_2 - y_1$ represents the change in the y-coordinate (vertical change or the "rise") while $x_2 - x_1$ in the denominator represents the change in the x-coordinate (horizontal change or the "run"). Hence, the slope is often referred to as the ratio of 'rise over run'.

Positive slopes indicate that the line is rising from left to right, negative slopes indicate a line falling from left to right, a zero slope indicates a horizontal line, and undefined slope indicates a vertical line.

> **Key Point**
>
> Slope characteristics: positive for upward tilt, negative for downward, zero for horizontal, and undefined for vertical lines.

Example Calculate the slope of the line passing through the points $A(2, 3)$ and $B(5, 11)$.

Solution: We use the formula for the slope: $m = \frac{y_2 - y_1}{x_2 - x_1}$. Plugging in our points A and B into this formula gives us: $m = \frac{11-3}{5-2} = \frac{8}{3}$. So, the slope of the line is $\frac{8}{3}$.

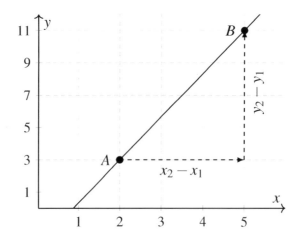

4.2 Formulating Linear Equations

A linear equation can be in many forms. In this topic, we will focus on the slope-intercept form, $y = mx + b$. m is the slope and b is the y-intercept, the point at which the line crosses the y-axis.

🔔 Key Point

A linear equation in slope-intercept form is given by $y = mx + b$ where m is the slope and b is the y-intercept.

If you know the slope and have a point on the line, determining the y-intercept is simple. Given a point $P(x_1, y_1)$ on the line and the slope m, the y-intercept b can be calculated using the following rearranged form of the equation $y = mx + b$:

🔔 Key Point

To find the y-intercept b of a line with a point $P(x_1, y_1)$, use the formula $b = y_1 - mx_1$.

📋 Example
Find the equation of the line that passes through the point $P(2, 3)$ with a slope of $\frac{1}{2}$.

Solution: We can find the y-intercept b using the formula: $b = y_1 - mx_1 = 3 - \frac{1}{2} \times 2 = 2$.

Therefore, the equation of the line is given by substituting m and b in the slope-intercept form: $y = mx + b = \frac{1}{2}x + 2$.

4.3 Deriving Equations from Graphs

We can derive the equation of a line from its graph by focusing on two key elements: the slope and the y-intercept.

> **Key Point**
>
> To turn a graph into an equation, we primarily need to identify two aspects – the slope and the y-intercept.

The slope is found by taking any two points on the line, and calculating the rise divided by the run. This represents the rate of change in the y values per unit change in the x values.

The y-intercept is simply the point where the line crosses the y-axis. This provides the constant term in the linear equation, denoted as 'b' in the standard form $y = mx + b$, where 'm' represents the slope.

> **Key Point**
>
> The y-intercept is the point where a line crosses the y-axis. In the equation $y = mx + b$, b represents this y-intercept, while m is the slope of the line.

Example Derive the equation of the line from the graph given in the following figure:

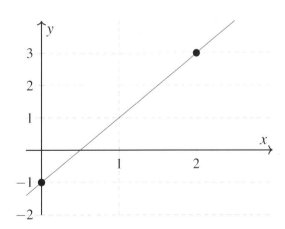

Solution: For this graph, we have two key points: $(0, -1)$ and $(2, 3)$.

Understanding that the y-intercept is the point where the line intercepts y-axis, we can see the y-intercept (b) is -1.

Next, finding the slope (m) by taking these two points gives us $\frac{3-(-1)}{2-0} = 2$. Thus, the derived equation from the graph is $y = 2x - 1$.

4.4 Understanding Slope-Intercept and Point-Slope Forms

A line can be represented in two forms: slope-intercept form and point-slope form.

> **Key Point**
>
> The Slope-Intercept Form: Represented as $y = mx + b$, where m is the slope and b is the y-intercept.

This form gives you a straightforward view of both the slope and the y-intercept of a line. which immediately exposes the rate of change and the starting point of the line on the y-axis.

> **Key Point**
>
> The Point-Slope Form: Represented as $y - y_1 = m(x - x_1)$, where m is the slope and (x_1, y_1) is a known point on the line.

This form is particularly helpful when we know the slope and at least one point that lies on the line. It emphasizes the relationship that every point has with each other on the line through the common slope.

Example Convert the equation $y = 3x - 2$ from slope-intercept form to point-slope form using the point $(2, 4)$ on the line.

Solution: From the given equation, we have that $m = 3$. For point-slope form, we have $y - y_1 = m(x - x_1)$, substituting $(x_1, y_1) = (2, 4)$ and $m = 3$ gives: $y - 4 = 3(x - 2)$.

Example Given that a line passes through the point $(1, 2)$ and has a slope of 3, write the equation in slope-intercept form and point-slope form.

Solution: If we denote the point as (x_1, y_1) and the slope as m, then for the point-slope form $y - y_1 = m(x - x_1)$, the equation of the line becomes: $y - 2 = 3(x - 1)$.

To express this in slope-intercept form, we rearrange the equation to the form $y = mx + b$: $y = 3x - 1$.

4.5 Writing Point-Slope Equations from Graphs

We can create equations from a line depicted graphically. The point-slope form is especially valuable in this regard as its format $(y - y_1 = m(x - x_1))$ only requires knowledge of any one point on the line (x_1, y_1), as well as the line's slope m.

Key Point

> To turn a graph into point-slope equation, identify a specific point and the slope. Plug in these values into the point-slope form $y - y_1 = m(x - x_1)$ to create your equation.

It must be noted that the choice of point does not affect the line that the equation will represent. Different points would result in different equations, but upon simplifying all these equations will yield the same equivalent line.

Example Find the equation of a graph line passing through the point $A(3,2)$ and having a slope of -2.

Solution: Identify the slope (m) and the point (x_1, y_1), then apply the point-slope form.

So, with $m = -2$, $x_1 = 3$, and $y_1 = 2$, we substitute these values into the point-slope form equation: $y - y_1 = m(x - x_1)$. This gives us: $y - 2 = -2(x - 3)$.

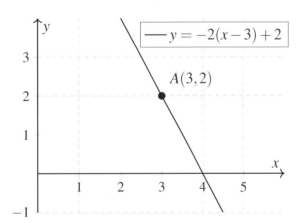

Example For a line on a graph passing through the point $P(4, -1)$ with a slope of 0.5, what will be the point-slope form equation?

Solution: In this case, we have the point $(x_1, y_1) = (4, -1)$ and the slope $m = 0.5$. Substitute these values into the point-slope form equation: $y - y_1 = m(x - x_1)$. Which results in: $y - (-1) = 0.5(x - 4)$ or simply: $y + 1 = 0.5(x - 4)$.

This equation represents, in the point-slope form, the line passing through point P with a slope of 0.5.

4.6 Identifying x- and y-intercepts

The x-intercept of a line refers to the point where it crosses the x-axis, while the y-intercept is where it crosses the y-axis. Knowing the intercepts allows us to quickly plot a graph, and they provide valuable insights into the behavior of the function.

Key Point

> The x- and y-intercepts are the points where the line crosses the x-axis and y-axis, respectively. To find the x-intercept, set $y = 0$ in the equation and solve for x. Similarly, to find the y-intercept, set $x = 0$ in the equation and solve for y.

Example Find the x- and y-intercepts of the line given by $2x - 3y = 6$.

Solution: To find the x-intercept, we set $y = 0$ in the equation and solve for x: $2x - 3(0) = 6 \Rightarrow 2x = 6 \Rightarrow x = 3$. So, the x-intercept is at the point $(3, 0)$.

Similarly, to find the y-intercept, we set $x = 0$: $2(0) - 3y = 6 \Rightarrow -3y = 6 \Rightarrow y = -2$. So, the y-intercept is at the point $(0, -2)$.

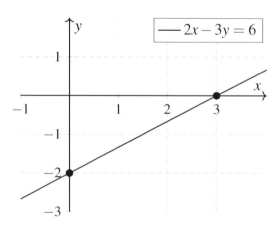

Example Identify the x- and y-intercepts for the line represented by the equation $5x = 7y$.

Solution: Setting $y = 0$ to find the x-intercept, we get: $5x = 7(0) \Rightarrow x = 0$. So, the x-intercept is at the point $(0,0)$.

Setting $x = 0$ to find the y-intercept, we also get: $5(0) = 7y \Rightarrow y = 0$. So, the y-intercept is also at the point $(0,0)$.

4.7 Graphing Standard Form Equations

The standard form of a linear equation is $Ax + By = C$, where A, B, and C are constants, and A and B are not both zero.

To graph an equation given in standard form, we can use the x-intercepts and y-intercepts. This approach is commonly known as the intercept method of graphing.

Key Point

The intercept method of graphing involves identifying and marking the x- and y-intercepts on the graph, then joining these points to form the line that represents the equation.

Example

Graph the standard form equation $3x - 2y = 6$.

Solution: First, we identify the x and y intercepts. Setting $y = 0$, solving for x we find the x-intercept is $x = 2$ (i.e., the point $(2,0)$).

Setting $x = 0$, solving for y, we find the y-intercept is $y = -3$ (i.e., the point $(0,-3)$).

We can now plot these two points on the graph and join them with a straight line. This line represents the equation $3x - 2y = 6$. See the figure below for the graph.

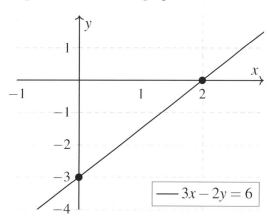

Example

Draw the graph for the equation $4x+2y=8$.

Solution: First step is to find the x-intercept. Setting $y=0$, we solve for x to obtain $x=2$. The x-intercept is therefore at the point $(2, 0)$.

Next, we find the y-intercept. Setting $x=0$, we solve for y to get $y=4$. The y-intercept is therefore at the point $(0, 4)$.

We now plot the points $(2,0)$ and $(0,4)$ on the graph and draw a straight line passing through them. This line represents the graph for the equation $4x+2y=8$.

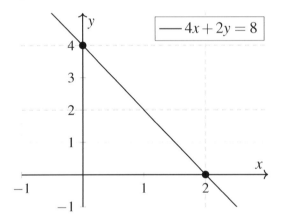

4.8 Understanding Horizontal and Vertical Lines

Although horizontal and vertical lines may seem much simpler than other linear functions, they have peculiar characteristics and definitions that need to be carefully noted.

Key Point

> A key characteristic of horizontal and vertical lines is that their slopes are unique. The slope of a horizontal line is always zero, while a vertical line does not have a finite slope, instead it is considered undefined.

In mathematical terms:

- For any horizontal line $y=b$, the slope $m=0$.

- For any vertical line $x=a$, the slope is ∞ (undefined).

This is a significant departure from other linear functions, as the slope is often a crucial part of the line's definition. For these lines we define them by their intercepts. The result is that vertical lines do not have a

4.9 Graphing Horizontal or Vertical Lines

y-intercept, and horizontal lines do not have an x-intercept unless they intersect the origin.

🔔 Key Point

Horizontal lines $y = b$ have no x-intercept, and vertical lines $x = a$ have no y-intercept, except when they pass through the origin.

 Determine whether the line given by the equation $x = 5$ is horizontal or vertical.

Solution: The given equation is in the form $x = a$, where a is a constant. Therefore, the line is a vertical line.

 Determine whether the line $y = 3$ is horizontal or vertical.

Solution: The given equation has the form $y = b$, where b is a constant. Consequently, the line is a horizontal line.

4.9 Graphing Horizontal or Vertical Lines

Graphing horizontal and vertical lines is straightforward.

🔔 Key Point

A horizontal line $y = b$, runs parallel to the x-axis, crossing at point b on the y-axis. A vertical line $x = a$, runs parallel to the y-axis, intersecting at point a on the x-axis.

To graph a horizontal or vertical line, we just locate the appropriate intercept on the correct axis and draw a straight line parallel to the other axis.

 Graph the line given by the equation $x = 3$.

Solution: The given equation is of the form $x = a$, indicating a vertical line. Therefore, we start by locating the value 3 on the x-axis. We draw a straight line parallel to the y-axis (vertical), passing through the point $(3, 0)$ on the x-axis.

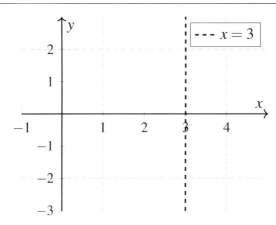

Example
Graph the line given by the equation $y = -2$.

Solution: The equation is of the form $y = b$, which indicates a horizontal line. Here, we begin by finding the value -2 on the y-axis. After that, we draw a straight line parallel to the x-axis (horizontal) passing through the point $(0, -2)$ on the y-axis. This is the graphical representation of the equation $y = -2$.

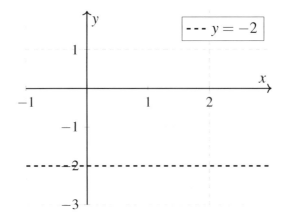

4.10 Graphing Point-Slope Form Equations

The point-slope form of a linear equation is given by $y - y_1 = m(x - x_1)$, where (x_1, y_1) is a point on the line and m is the slope of the line. With knowledge of a single point and the slope, we can graph a line.

Key Point

For plotting slope-intercept form, we begin by first plotting the known point on the graph. We then use the slope to identify a second point and draw a line through these two points.

4.10 Graphing Point-Slope Form Equations

Example Graph the line for the equation $y - 2 = 3(x + 1)$.

Solution: This equation is already in point-slope form. The point $(-1, 2)$ and the slope 3 are given.

1. Start by plotting the point $(-1, 2)$ on the Cartesian plane.

2. The slope is 3, which can be written as $\frac{3}{1}$. Remember that the numerator (3) represents the vertical change (rise) and the denominator (1) represents the horizontal change (run).

3. From the point $(-1, 2)$, move 3 units up (rise) and 1 unit to the right (run), obtaining a new point $(0, 5)$.

4. Draw a straight line passing through these two points. This line represents the equation $y - 2 = 3(x + 1)$.

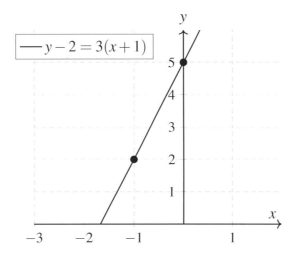

Example Graph the line of the equation $y - 4 = -2(x - 3)$.

Solution: The point and the slope provided are $(3, 4)$ and -2, respectively.

1. Initially, plot the point $(3, 4)$ on the graph.

2. The slope -2 can be expressed as $\frac{-2}{1}$.

3. Starting from the point $(3, 4)$, move 2 units down (rise, as it is negative) and 1 unit to the right (run), obtaining a new point $(4, 2)$.

4. Draw a straight line passing through these two points. The line represents the equation $y - 4 = -2(x - 3)$.

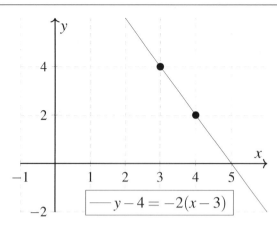

4.11 Understanding Parallel and Perpendicular Lines

In the Cartesian plane, two lines are said to be parallel if they have the same slope and they do not intersect at any point. This results from the fact that they maintain a constant distance from each other. As they have the same slope, the lines $y = m_1 x + c_1$ and $y = m_2 x + c_2$ are parallel if $m_1 = m_2$ and $c_1 \neq c_2$.

Key Point

Two lines are parallel if and only if their slopes are equal.

On the other hand, two lines are said to be perpendicular if their slopes are the negative reciprocal of each other and they intersect at a right angle (90-degree angle). Meaning, if the line 1 has a slope m_1, line 2 will be perpendicular to line 1 if its slope $m_2 = -\frac{1}{m_1}$.

Key Point

Two lines are perpendicular if and only if their slopes are negative reciprocals.

This understanding is essential in scenarios where you are required to find lines parallel or perpendicular to a given line.

Example Given that the equation of line L_1 is $y = 2x - 3$, find the equation of line L_2 passing through the point $(3,1)$ and is parallel to L_1.

Solution: We know from the equation of L_1 that its slope $m_1 = 2$.

Since L_2 is parallel to L_1, it must have the same slope. Hence, the slope of L_2, m_2, is also 2.

The point-slope form of a line is $y - y_1 = m(x - x_1)$, where m is the slope, and (x_1, y_1) is a point on the line.

So, plugging 2 for m, 3 for x_1, and 1 for y_1 in the equation, we get: $y - 1 = 2(x - 3)$ or $y = 2x - 5$.

So, the equation of L_2 is $y = 2x - 5$.

Example Given that the equation of line L_1 is $y = -3x + 2$, find the equation of line L_2 that passes through the point $(-1, 4)$ and is perpendicular to L_1.

Solution: We know from the equation of L_1 that its slope $m_1 = -3$.

Since L_2 is perpendicular to L_1, its slope m_2 must be the negative reciprocal of m_1. Hence, $m_2 = -\frac{1}{m_1} = -\frac{1}{-3} = \frac{1}{3}$.

The point-slope form of a line is $y - y_1 = m(x - x_1)$, where m is the slope and (x_1, y_1) is a point on the line.

So, plugging $\frac{1}{3}$ for m, -1 for x_1, and 4 for y_1 in the equation, we get: $y - 4 = \frac{1}{3}(x + 1)$ or $y = \frac{1}{3}x + \frac{13}{3}$.

So, the equation of L_2 is $y = \frac{1}{3}x + \frac{13}{3}$.

4.12 Comparing Linear Function Graphs

There are typically two primary features we are interested in when comparing linear function graphs: the slope, and the y-intercept.

The slope of a linear function describes the steepness of the line, and it also indicates the direction in which the line is moving: upward or downward. If two lines have the same slope, they are parallel to each other on the graph.

Comparing the y-intercepts of linear functions can tell us where each line crosses the y-axis. If two lines intersect at a common point that is not on the y-axis, their y-intercepts will be different. If two lines cross on the y-axis, meaning they have common y-intercepts, it does not guarantee that those lines are the same, but it does mean that they cross at the same point on the y-axis.

Key Point

A comparison of slopes can show whether lines are parallel, while a comparison of y-intercepts can show where each line crosses the y-axis.

Example Given two linear equations: $y = 2x - 1$ (equation 1) and $y = 2x + 3$ (equation 2), identify the similarities and differences between these two linear functions.

Solution: Analyzing these two equations, we find that both equations have the same slope of 2. Meaning the lines represented by these equations are parallel to each other.

However, they have different y-intercepts: -1 for equation 1 and $+3$ for equation 2. This means that each line crosses the y-axis at a different point. So, while the lines are parallel, they are not identical because they cross the y-axis at different points.

Example Given two linear equations: $y = -x + 2$ (equation 3) and $y = x - 2$ (equation 4), what are the similarities and differences between these two linear functions?

Solution: In this case, equation 3 and equation 4 have differing slopes: -1 for equation 3 and 1 for equation 4. This indicates that the two lines are not parallel and they will intersect at some point on the graph.

Additionally, the y-intercepts are different. The y-intercept for equation 3 is $+2$ while it is -2 for equation 4. This tells us that equation 3 and equation 4 cross the y-axis at different points, further reinforcing the fact that they are distinct lines.

The slopes being negative reciprocals (-1 and 1) prove the lines are perpendicular, intersecting at a right angle.

4.13 Graphing Absolute Value Equations

Absolute value equations have a unique characteristic when graphed: they form a "V" or inverted "V" shape. This characteristic is due to the property of the absolute value, which states that the result of an absolute value operation is always positive, or zero.

The standard form of absolute value equation is $y = |x|$, which produces a "V" shape on the graph. When we add any number (except zero) to x or y the vertex of the equation moves to different positions in the graph.

An important aspect of graphing absolute value equations is understanding how these modifications to the equation affect the graph.

4.13 Graphing Absolute Value Equations

Key Point

The graph of an absolute value equation forms a "V" or inverted "V" shape. The vertex of the graph changes position based on modifications to the equation.

Example
Graph the equation $y = |x|$.

Solution: This is the standard form of an absolute value equation. To graph it, we plot points based on input and output values:

If we take x-values $-2, -1, 0, 1, 2$, then we calculate the corresponding y-values as the absolute values of these x-values, so we will have y-values as $2, 1, 0, 1, 2$.

Plotting these points $(-2,2), (-1,1), (0,0), (1,1), (2,2)$ on a Cartesian plane, and connecting the plotted points, we will get a "V" shape. The vertex (lowest point) of this "V" shape is $(0,0)$. See the following figure for a visual representation of the graph.

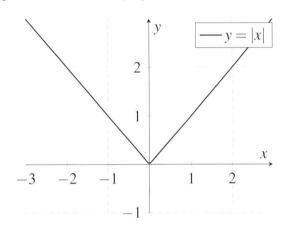

Example
Graph the equation $y = |x - 2|$.

Solution: Here, instead of the absolute value of x, we have the absolute value of $(x - 2)$. This moves the vertex of the "V" shape to the right by 2 units:

If we use x-values $0, 1, 2, 3, 4$, the corresponding y-values will be $2, 1, 0, 1, 2$ respectively.

Plotting these points $(0,2), (1,1), (2,0), (3,1), (4,2)$ on a Cartesian plane and connecting them, a "V" shape is formed with the vertex at $(2,0)$.

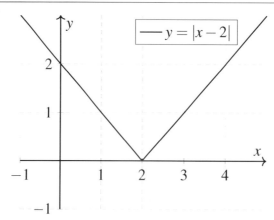

4.14 Solving Two-Variable Word Problems

Word problems involving two variables are everyday scenarios that require setting up two equations to solve for the two unknowns. An understanding of linear functions, graphing, and equations is necessary to solve these problems.

Key Point

The key to solving two-variable word problems lies in identifying the two unknowns, representing them as variables, and setting up the correct equations reflecting the problem's requirements.

Example A store sold 50 items. Each T-shirt costs $10 and each cap costs $15. The total income is $550. How many T-shirts and how many caps were sold?

Solution: Let's denote the number of T-shirts sold as x and the number of caps as y. We can then form two equations based on the problem:

The first equation comes from the total number of items: $x + y = 50$.

The second equation represents the total income: $10x + 15y = 550$.

We can now solve this system of equations. Subtracting 10 times the first equation from the second equation, we get $5y = 50$, which means $y = 10$. Substituting $y = 10$ into the first equation, we get $x = 40$.

Therefore, the store sold 40 T-shirts and 10 caps.

Example A school has 200 students, some in class A, rest in class B. Class A has 20 students more than class B. How many students are there in each class?

Solution: Let's assign the variable x to students in class B, then class A will have $x+20$ students.

The equations will be: $x+(x+20) = 200$.

Solving this equation, we get $2x = 180$, so $x = 90$. That means there are 90 students in class B and $90+20 = 110$ students in class A.

4.15 Practices

Solve:

1) Calculate the slope of the line passing through the points $A(-3,1)$ and $B(4,-1)$.

2) Calculate the slope of the line passing through the points $A(0,0)$ and $B(5,15)$.

3) Calculate the slope of the line passing through the points $A(-2,6)$ and $B(4,-8)$.

4) Calculate the slope of the line passing through the points $A(-1,3)$ and $B(6,-10)$.

5) Calculate the slope of the line passing through the points $A(2,-5)$ and $B(7,10)$.

Fill in the blank:

6) The form of a linear equation $y = mx+b$ is known as the _____ form.

7) In the equation $y = mx+b$, the 'm' stands for _____.

8) In the equation $y = mx+b$, the 'b' is the _____ of the line.

9) If a line passes through the point $P(x_1, y_1)$ and has a slope m, the y-intercept 'b' can be calculated as $b = y_1 -$ _____.

10) If a line has slope zero, it is a _____ line.

Solve:

11) Write the equation of the line in slope-intercept form given that the slope is 4 and the y-intercept is -2.

12) Convert the equation $y = 2x + 5$ originally in slope-intercept form, to point-slope form using the point (1, 7) on the line.

13) A line passes through the point $(5, -1)$ and has a slope of -3. Write the equation in point-slope form.

14) Given the line $y - 2 = 4(x - 3)$ in point-slope form, convert it to slope-intercept form.

15) A line has the equation $y = -x + 6$. What is the slope and the y-intercept of this line?

Fill in the Blank:

16) Given the equation $4x = 5y$, the x-intercept is _____ and the y-intercept is _____.

17) For the equation $6x - 2y = 12$, the x-intercept is _____ and the y-intercept is _____.

18) If the equation of the line is $7x + 4y = 28$, the x-intercept and y-intercept are _____ and _____ respectively.

19) If the linear equation is $8x - 3y = 24$, the x-intercept is _____ and the y-intercept is _____.

20) Give the y-intercept and x-intercept for the equation $9x + 3y = 18$.

Fill in the Blank:

21) The slope of a horizontal line is always _____.

22) The slope of a vertical line is always _____.

23) The equation of a vertical line is of the form _____.

24) The equation of a horizontal line is of the form _____.

25) A vertical line does not have a _____.

Solve:

4.15 Practices

26) Given that the equation of line L_1 is $y = 4x - 7$, find the equation of line L_2 passing through the point $(2, 1)$ and is parallel to L_1.

27) Given that the equation of line L_1 is $y = -5x + 2$, find the equation of line L_2 that passes through the point $(-1, -4)$ and is perpendicular to L_1.

28) Given the line $y = 3x - 1$, what is the equation of a line parallel to this line that passes through the point $(1, 4)$?

29) Given that the equation of line L_1 is $y = -4x - 2$, find the equation of line L_2 that passes through the point $(1, -3)$ and is perpendicular to L_1.

30) Given the line $y = -2x + 3$, what is the equation of a line perpendicular to this line that passes through the point $(-2, -1)$?

True or False:

31) If two lines are parallel, they must have the same y-intercept.

32) The y-intercept tells where a line crosses the y-axis.

33) If two lines intersect at a common point that is not on the y-axis, they must have the same slope.

34) If a line has a negative slope, it is moving downward on the graph.

35) If two lines have the same slope and the same y-intercept, they are the same line.

Answer Keys

1) $m = -\frac{2}{7}$
2) $m = 3$
3) $m = -\frac{7}{3}$
4) $m = -\frac{13}{7}$
5) $m = 3$
6) slope-intercept
7) slope
8) y-intercept
9) mx_1
10) horizontal
11) $y = 4x - 2$
12) $y - 7 = 2(x - 1)$
13) $y + 1 = -3(x - 5)$
14) $y = 4x - 10$
15) Slope $= -1$, y-intercept $= 6$
16) x-intercept:$(0,0)$, y-intercept:$(0,0)$
17) x-intercept:$(2,0)$, y-intercept:$(0,-6)$
18) x-intercept:$(4,0)$, y-intercept:$(0,7)$
19) x-intercept:$(3,0)$, y-intercept:$(0,-8)$
20) x-intercept:$(2,0)$, y-intercept:$(0,6)$
21) 0
22) Undefined
23) $x = a$
24) $y = b$
25) y-intercept
26) $y = 4x - 7$
27) $y = \frac{1}{5}x - \frac{19}{5}$
28) $y = 3x + 1$
29) $y = \frac{1}{4}x - \frac{13}{4}$
30) $y = \frac{1}{2}x$
31) False
32) True
33) False
34) True
35) True

Answers with Explanation

1) Using the slope formula, we get: $m = \frac{-1-1}{4-(-3)} = -\frac{2}{7}$.

2) Using the slope formula, we get: $m = \frac{15-0}{5-0} = 3$.

3) Using the slope formula, we get: $m = \frac{-8-6}{4-(-2)} = -\frac{7}{3}$.

4) Using the slope formula, we get: $m = \frac{-10-3}{6-(-1)} = -\frac{13}{7}$.

5) Using the slope formula, we get: $m = \frac{10-(-5)}{7-2} = 3$.

6) The linear equation $y = mx + b$ is in slope-intercept form.

7) In the equation $y = mx + b$, the 'm' is the slope of the line.

8) In the equation $y = mx + b$, the 'b' is the y-intercept of the line.

9) Using $b = y_1 - mx_1$, we can calculate 'b', the y-intercept of the line.

10) A line that has slope zero is a horizontal line.

11) By substituting $m = 4$ for slope and $b = -2$ into the slope-intercept form $y = mx + b$, we get $y = 4x - 2$.

12) From the given equation, we have $m = 2$. Substituting $(x_1, y_1) = (1, 7)$ and $m = 2$ into point-slope form, we get $y - 7 = 2(x - 1)$.

13) Given $(x_1, y_1) = (5, -1)$ and $m = -3$, in point-slope form, the equation of the line becomes $y + 1 = -3(x - 5)$.

14) By distributing the number 4 in the equation and shifting -2 to the other side, we get the slope-intercept form $y = 4x - 10$.

15) In the slope-intercept form of the line $y = mx + b$, the coefficient of x is the slope, and the constant term is the y-intercept. Therefore, from the equation $y = -x + 6$, the slope (m) is -1, and the y-intercept (b) is 6.

16) Setting $y = 0$ and $x = 0$ to find the x- and y-intercepts respectively, we get (0,0) for both.

17) To find x-intercept, when $y = 0$, $x = 2$. For y-intercept, when $x = 0$, $y = -6$.

18) The x-intercept (when $y = 0$) is $x = 4$ and the y-intercept (when $x = 0$) is $y = 7$.

19) x-intercept (when $y = 0$) is $x = 3$ and the y-intercept (when $x = 0$) is $y = -8$.

20) By setting $y = 0$ and $x = 0$, we find the x-intercept (2,0) and the y-intercept (0,6) respectively.

21) A characteristic of horizontal lines is that their slope (rate of change) is always zero as they run parallel to the x-axis.

22) The slope of a vertical line is undefined because the line runs parallel to the y-axis. It does not have a distinct rate of change.

23) The standard form of the equation of a vertical line is $x = a$, where a is the x-coordinate of any point on the line.

24) The standard form of the equation of a horizontal line is $y = b$, where b is the y-coordinate of any point on the line.

25) Vertical lines are parallel to the y-axis and thus do not have a y-intercept.

26) Since L_2 is parallel to L_1, it must have the same slope. Hence, the equation of L_2 is $y - 1 = 4(x - 2)$, simplifying this will give the equation of L_2 is $y = 4x - 7$.

27) Since L_2 is perpendicular to L_1, its slope m_2 must be the negative reciprocal of m_1 (which is -5). Hence, $m_2 = -\frac{1}{m_1} = -\frac{1}{-5} = \frac{1}{5}$. Plugging in the values into the point-slope form will give the equation: $y - (-4) = \frac{1}{5}(x - (-1))$, simplifying which will give the equation of L_2 is $y = \frac{1}{5}x - \frac{19}{5}$.

28) Since the slopes of parallel lines are equal, the slope of the new line is 3. Substituting the slope and the point into the point-slope form will give $y - 4 = 3(x - 1)$. Simplifying this will give the equation of the new line as $y = 3x + 1$.

29) Since L_2 is perpendicular to L_1, its slope m_2 must be the negative reciprocal of m_1 (which is -4). Hence, $m_2 = -\frac{1}{m_1} = \frac{1}{4}$. Plugging in the values into the point-slope form will give the equation: $y - (-3) = \frac{1}{4}(x - 1)$, simplifying which will give the equation of L_2 is $y = \frac{1}{4}x - \frac{13}{4}$.

30) The slope of a line perpendicular to the given line is the negative reciprocal of its slope, so the new line has a slope of $\frac{1}{2}$. Putting the slope and the point into the point-slope form will give $y - (-1) = \frac{1}{2}(x - (-2))$. Simplifying this will give the equation of the new line as $y = \frac{1}{2}x$.

31) Two lines can be parallel (have the same slope) and still have different y-intercepts.

32) The y-intercept indeed shows the point where a line crosses the y-axis.

33) If two lines intersect at a common point, they have different slopes. Lines with the same slope are parallel and never intersect.

34) A negative slope means a line is moving downward from left to right on the graph.

35) If two lines have the same slope and y-intercept, they will coincide with each other on the graph and thus, represent the same line.

5. Inequalities and Systems of Equations

5.1 Solving One-Step Inequalities

An inequality compares two values. Inequalities could be expressed mathematically as less than, greater than, less than or equal to, and greater than or equal to. For instance, $x > 5$ is an inequality and it states that x is greater than 5.

One-step inequalities are inequalities that can be solved through one single operation: addition, subtraction, multiplication, or division. The primary goal while dealing with one-step inequalities is to isolate the variable, i.e., to get the variable on one side of the inequality.

🔔 Key Point

Keep in mind to change the direction of the inequality sign when you multiply or divide both sides by a negative number. This is a crucial factor while working with inequalities.

 Solve the inequality $x + 6 > 11$.

Solution: To isolate the variable x, subtract 6 from both sides: $x + 6 - 6 > 11 - 6$. You will then get $x > 5$. Therefore, all numbers greater than 5 are solutions of the inequality.

 Solve the inequality $-3y < 15$.

Solution: Multiply both sides by $-\frac{1}{3}$ to isolate the variable y. Don't forget to flip the inequality sign because we divided by a negative number. So we have: $y > -5$. Therefore, all numbers greater than -5 are solutions of the inequality.

 **Example** If $m - 7 \leq 5$, find m.

Solution: Isolate m by adding 7 to both sides: $m \leq 12$. Therefore, all numbers less than or equal to 12 are solutions of the inequality.

5.2 Solving Multi-Step Inequalities

Multi-Step Inequalities are more complex than one-step inequalities, but the core concepts remain the same. The primary goal remains isolating the variable, but unlike one-step inequalities, this process may require more than one mathematical operation.

Just like in one-step inequalities, be cautious while dividing or multiplying by a negative inequality, the inequality symbol must be reversed.

Key Point

Regardless of the complexity of the inequality, always prioritize isolating the variable. This often simplifies the problem significantly, and you can often break down the problem into several smaller one-step inequalities.

 **Example** Solve the inequality $2x - 5 > 3$.

Solution: In order to isolate x, you should first add 5 to both sides of the inequality. This gives us: $2x > 3 + 5$. which simplifies to: $2x > 8$. Then, divide both sides by 2 to fully isolate x, giving: $x > 4$. Therefore, all numbers greater than 4 are solutions of the inequality.

 **Example** Solve the inequality $3x + 5 \leq 17$.

Solution: Start by subtracting 5 from both sides to isolate the term with the variable: $3x \leq 17 - 5$. Which simplifies to: $3x \leq 12$. Then divide both sides by 3: $x \leq 4$. So, the solution to the inequality is all numbers less than or equal to 4.

5.3 Working with Compound Inequalities

A compound inequality involves at least two inequalities that are joined together by the terms "and" or "or".

> **Key Point**
>
> As a rule, *'and'* typically narrows down the solution set (think intersection) while *'or'* tends to expand the solution set (union).

 Solve the compound inequality $x > 2$ and $x \leq 5$.

Solution: The inequality says x is greater than 2 *and* x is less than or equal to 5. This corresponds to a range of solutions. As such, the solution set is $2 < x \leq 5$.

To graph this, draw a line from 2 (not included, so we represent this with an open circle) to 5 (included, so we draw a filled circle) on the number line.

 Solve the compound inequality $x < -1$ or $x \geq 4$.

Solution: The inequality represents two sets of values for x, one where x is less than -1, and the other where x is greater than or equal to 4. Its solution is the union of these two sets.

When graphing, plot all the points less than -1 and points greater than or equal to 4. Use open circle for -1 and filled circle for 4.

5.4 Graphing Solutions to Linear Inequalities

To graph a linear inequality, we start by graphing the corresponding linear equation ('boundary line'), changing it into an equal sign. Once the line is graphed, we decide whether it is solid (for "$\leq$" or "$\geq$") or dashed (for "$<$" or "$>$").

> **Key Point**
>
> A solid line includes values on the line as solutions, while a dashed line excludes those values.

After graphing the boundary line, we choose a test point, usually $(0,0)$ unless the line goes through the origin. If this point satisfies the inequality, shade the area that includes the test point. Otherwise, shade the other side.

5.4 Graphing Solutions to Linear Inequalities

🔔 Key Point

Use $(0,0)$ as the test point (unless it is on the inequality line) and shade the region where this point satisfies the inequality.

📋 Example
Graph the inequality $y \geq 2x + 3$.

Solution: The boundary line is $y = 2x + 3$, which has a y-intercept of 3 and a slope of 2. We plot the line as solid, because of the "$\geq$" sign.

For the shading, choose a test point, say $(0,0)$. Substituting into our inequality, we get $0 \geq 2(0) + 3$, which is false. Hence, we shade the region not containing $(0,0)$, which is above the line.

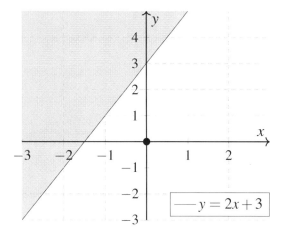

📋 Example
Graph the inequality $y < -x + 1$.

Solution: The equation of the boundary line is $y = -x + 1$, with a y-intercept of 1 and slope of -1. We draw this as a dashed line because of the "$<$" sign.

Choosing the test point $(0,0)$, the inequality becomes $0 < -0 + 1$, which is true. Therefore, we shade the region containing the test point, which is below the dashed line.

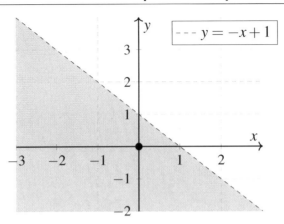

5.5 Writing Linear Inequalities from Graphs

Graphs show us a geometric representation of an inequality. Our goal is to derive the algebraic inequality equation or inequality expression by examining this graphical representation.

The plotted line indicates the linear equation, either solid or dashed. The shaded region, on the other hand, reflects the inequality part as it represents all possible solutions. The solid line tells us that the points on the line are part of the solution ($\leq$ or $\geq$), whereas a dashed line excludes the points on the line from the solution ($>$ or $<$).

Key Point

A dashed line in a graph signifies "<" or ">", while a solid line signifies "$\leq$" or "$\geq$".

Before identifying the line's equation, it is essential first to decide whether it is solid or dashed. Next, you can find the equation by identifying the slope and y-intercept.

Example Consider a solid line graph passing through $(0, 0)$ and $(1, 3)$. The shaded region lies above the line. Write the corresponding inequality.

Solution: Since the line is solid, the inequality symbol will be either "$\leq$" or "$\geq$". As the shaded region lies above the line, we know our inequality symbol must be "$\geq$".

The line passes through the origin $(0, 0)$, so the y-intercept is 0. The slope can be determined from the two points, $\frac{y_2-y_1}{x_2-x_1} = \frac{3-0}{1-0} = 3$. Hence, the inequality will be $y \geq 3x$.

Example Imagine a dashed line graph passing through the points $(0, 2)$ and $(2, 1)$. The shaded area is below the line. Write the corresponding inequality.

Solution: Since the line is dashed, the inequality symbol will be either "<" or ">". As the shaded region is below the line, we know that our inequality symbol is "<".

The line passes through the point (0, 2), so the y-intercept is 2. The slope is calculated from the two points, $\frac{y_2-y_1}{x_2-x_1} = \frac{1-2}{2-0} = -\frac{1}{2}$. Therefore the inequality is $y < -\frac{1}{2}x + 2$.

5.6 Solving Advanced Linear Inequalities in Two Variables

Advanced linear inequalities involving two variables are essentially systems of inequalities which simultaneously govern two variables. Unlike simple linear inequalities, these involve two variables and their solutions generally represent a region in the coordinate plane, not just a line.

Advanced linear inequalities in two variables are systems of inequalities where solutions make all inequalities true simultaneously.

To solve such inequalities, we graph all inequalities in the system and find the intersecting region that makes all inequalities true. The line plotted could be dashed or solid, depending on the type of inequality.

 Consider the advanced linear inequality system:

$$\begin{cases} y \geq x+2 \\ y < -2x+1 \end{cases}$$

Solve and graph the system, finding the solution region.

Solution: Start by graphing the line $y = x+2$ as a solid line, since the inequality is $\geq$. Similarly, graph $y = -2x+1$ as a dashed line, due to the < inequality.

Choose a test point, say $(0,0)$, and substitute it into both inequalities:

- For $y \geq x+2$, we get $0 \geq 2$, which is false. So shade the region not containing the test point i.e., above the solid line.

- For $y < -2x+1$, we get $0 < 1$, which is true. So, we shade the area that includes $(0,0)$, i.e., below the dashed line.

The solution region, therefore, would be below the dashed line and above the solid line. It is the overlapping area where all the conditions of the inequalities are satisfied.

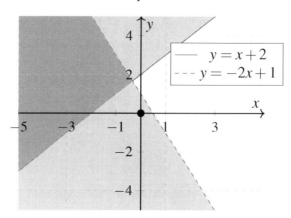

5.7 Graphing Solutions to Advanced Linear Inequalities

Let's focus on how to graphically represent the coordinates which satisfy advanced linear inequalities in two variables.

The solution to an advanced linear inequality in two variables is represented as a region on the coordinate plane, marked by multiple lines of inequality.

 Consider the following system of advanced linear inequalities:

$$\begin{cases} y \geq -x+3 \\ y < 2x+1 \end{cases}$$

Sketch a graph based on these inequalities, clearly identifying the solution region.

Solution: Start by graphing each inequality as if it were an equation: $y = -x+3$ is a downward sloping line that intercepts the y-axis at 3, and $y = 2x+1$ is an upward sloping line with y-intercept at 1.

In terms of shading, since $y \geq -x+3$, we shade everything above the line, and since $y < 2x+1$, we shade below the dashed line.

The solution to our system of inequalities occurs in the region where both shading overlap. This is the area that satisfies both inequalities simultaneously. See the following figure. The solution region is the overlapping area shaded in darker gray.

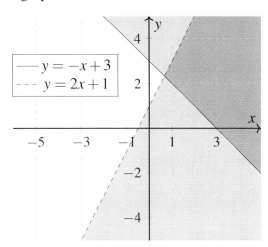

5.8 Solving Absolute Value Inequalities

To solve absolute value inequalities, we follow two broad types. One type includes those inequalities that are 'greater than' (or 'greater than or equal to'), and the other type includes those that are 'less than' (or 'less than or equal to').

> Absolute value inequalities bring in a layer of complexity which we resolve by transforming them into a system of inequalities without any absolute value.

It is crucial to note that for 'less than' absolute value inequalities, the solution set will usually involve an 'and' statement (or solutions between two values), while 'greater than' absolute value inequalities generally produce an 'or' statement (or solutions not within certain values).

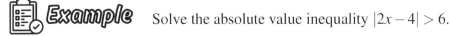 Solve the absolute value inequality $|2x - 4| > 6$.

Solution: Firstly, express the absolute value inequality as two separate inequalities: $2x - 4 > 6$ and $2x - 4 < -6$.

Solving each of these will give:

From $2x - 4 > 6$ we get $2x > 10$, thus $x > 5$.

From $2x-4 < -6$, we get $2x < -2$, thus $x < -1$.

These are the solution ranges for the absolute value inequality.

 Determine the range of values of x for the absolute value inequality $|3x+2| \leq 7$.

Solution: To solve this, express it as two separate inequalities:

$3x+2 \leq 7$ and $3x+2 \geq -7$.

Solving each of these will give:

From $3x+2 \leq 7$ we get $3x \leq 5$, thus $x \leq \frac{5}{3}$.

From $3x+2 \geq -7$, we get $3x \geq -9$, thus $x \geq -3$.

Therefore, our solution is $-3 \leq x \leq \frac{5}{3}$, written using an 'and' statement as we predicted.

5.9 Understanding Systems of Equations

A system of equations is a set of multiple linked linear equations. These equations share common variables, and the solutions to each equation must satisfy all other equations in the set.

> A solution to a system of equations is a set of variable values that satisfy all the equations in the system simultaneously.

The key in understanding a system of equations is recognizing that each equation provides a piece of information about the potential solutions. When viewed together, these equations create a system that can provide specific numerical values for each of the variables.

The simplest form of a system of equations involves two equations with two variables. Often, a system of equations is represented graphically, with each equation being shown as a straight line on a graph. The points where the lines intersect correspond to the solutions of the system.

 Consider the system of equations:

$$\begin{cases} x+y=7 \\ -x+y=1 \end{cases}$$

What is a solution to this system of equations?

Solution: In this case, we can solve for one variable in terms of the other using the first equation: $y = 7 - x$. And then substitute this into the second equation: $-x + 7 - x = 1$. Simplifying, we find: $2x = 6$. So $x = 3$. Finally, substituting $x = 3$ into the first equation yields $y = 4$: $3 + 4 = 7$

Therefore, the solution to this system of equations is $x = 3$ and $y = 4$.

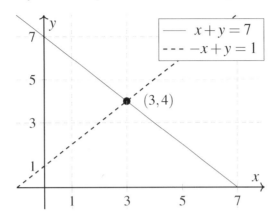

5.10 Determining the Number of Solutions to Linear Equations

When solving systems of equations, it is important to recognize the potential nature of solutions. These systems can yield one unique solution, infinitely many solutions, or no solution. The differentiating factor primarily revolves around the relationship between the equations involved in the system.

> **Key Point**
>
> There are three potential outcomes when solving systems of linear equations: one unique solution, infinitely many solutions, and no solution.

In the case of one unique solution, the lines represented by the equations intersect at a single point. When a system of equations has infinitely many solutions, the equations represent the same line. In such instances, every point on the line is a solution.

A system has no solution when the lines from the equations are parallel to each other. This means they never intersect, and there is not a set of values that satisfy all the equations simultaneously.

Key Point

To determine the number of solutions without solving the system, check the coefficients and constants. If the ratios of the coefficients of x, y, and the constants are equal, the system has infinite solutions. If the coefficients are proportionally equal except for the constants, then there is no solution. Otherwise, the system has only one solution.

Example

Determine the number of solutions in the system of equations:

$$\begin{cases} 2x + 3y = 12 \\ 4x + 6y = 26 \end{cases}$$

Solution:

In this case, we compare coefficients and constants ratios:

The coefficient of x in the first equation to x in the second is $\frac{2}{4} = 0.5$

The coefficient of y in the first to y in the second is $\frac{3}{6} = 0.5$

The constant in the first equation to the constant in the second is $\frac{12}{26} \approx 0.46$

As they are not all equal, the system has no solutions. The lines would be parallel and never intersect.

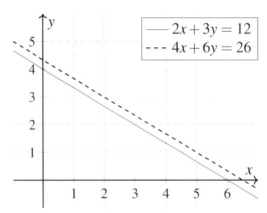

5.11 Writing Systems of Equations from Graphs

Creating a system of equations from graphs involves interpreting the relationships depicted visually on a coordinate plane and then representing these relationships using mathematical notation. This usually entails writing equations for two or more lines that intersect at a specific point.

5.11 Writing Systems of Equations from Graphs

Key Point

Looking at a graph, one can write down a system of equations by deciding the equations representing the lines and figuring out the intersection points.

The process of formulating systems of equations from graphs can be categorized into three primary steps:
1. Identify the slopes and y-intercepts of each line.
2. Write the equation of each line using the slope-intercept form $y = mx + b$, where m is the slope and b is the y-intercept.
3. Identify the intersection point(s), which is the solution to the system of equations.

Example

A graph contains two line described as follows:

Line A: It passes through the points $(0, 1)$ and $(4, 5)$

Line B: It passes through the point $(2, 3)$ and has slope of -1

Write a system of equations that represent the two lines.

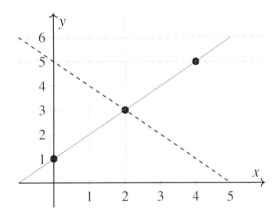

Solution:

For line A:

The slope is given by $m = \frac{y_2 - y_1}{x_2 - x_1} = \frac{5-1}{4-0} = 1$. The y-intercept (the point where the line crosses the y-axis) is given as $(0, 1)$, hence $b = 1$. Now we can write the equation for line A: $y = 1x + 1 = x + 1$.

For line B:

We already know the slope $m = -1$ and we have the point $(2, 3)$. To find b, we insert the given point and slope into the slope-intercept equation. Solving $y = mx + b$ gives $3 = -1 \times 2 + b$, which yields $b = 3 + 2 = 5$. The equation for line B can now be written as $y = -1x + 5 = -x + 5$.

Therefore, our system of equations based on the graph is:

$$\begin{cases} y = x+1 \\ y = -x+5 \end{cases} \implies \begin{cases} y - x = 1 \\ y + x = 5 \end{cases}$$

5.12 Solving Systems of Equations Word Problems

A key to solving word problems is having the ability to interpret the problem and translate it into mathematical language.

> **Key Point**
>
> To tackle system of equations word problems, we translate the problem's text into a system of equations, solve it, and then interpret the solution within the context of the problem.

Methods of Solving Systems of Equations There are two commonly used methods to solve systems of equations: the *substitution method* and the *elimination method*. In the substitution method, we solve one equation for one variable in terms of the other variable(s), and then substitute this expression into another equation to find the value of the remaining variable(s). In the elimination method, we add or subtract equations to eliminate one variable, making it easier to find the value of the remaining variable(s).

Example Billy and Ann are selling lemonade. Billy sells 1 cup of lemonade for $3. Ann sells 1 cup for $4. They want to sell a total of 35 cups for $115. How many cups does each person need to sell?

Solution: Let's denote by x the number of cups Billy sells and by y the number of cups Ann sells.

Then we have the following system of equations: $\begin{cases} x+y = 35 \\ 3x+4y = 115 \end{cases}$.

We can solve this system using the method of substitution or elimination. Here, we will use elimination.

Multiply the first equation by 3: $\begin{cases} 3x+3y = 105 \\ 3x+4y = 115 \end{cases}$.

Subtract the first new equation from the second: $4y - 3y = 115 - 105 \Rightarrow y = 10$.

Substitute $y = 10$ back into the first original equation: $x + 10 = 35 \Rightarrow x = 25$.

Therefore, Billy needs to sell 25 cups of lemonade, and Ann needs to sell 10 cups.

5.13 Solving Word Problems Involving Linear Equations

Sometimes you need to translate the problem to a linear equation, solve it, interpret the results, and ensure they make sense in the problem's context.

Key Point

> Solving a word problem involving linear equations entails translating the problem into a linear equation, solving the equation, and interpreting the result in the problem's real-world context.

Linear equations can represent a range of practical situations - from simple shopping scenarios to complex mixtures problems. Bear in mind, the terms used in word problems often represent mathematical operations. For example, the word 'total' usually means 'sum up', and 'per' usually indicates a division.

Example Chloe buys 5 apples and 6 bananas for $14. If the price of apples and bananas is the same, what is the price of an apple?

Solution: Let's denote by A the price of an apple and by B the price of a banana.

From the problem, we know that $A = B$ and that 5 apples and 6 bananas cost $14. Therefore, we can write the following equation: $5A + 6B = 14$. Since $A = B$, we replace B with A in the equation to get: $5A + 6A = 14 \Rightarrow 11A = 14$.

To find the price of an apple, we solve the equation for A: $A = \frac{14}{11} \approx \$1.27$. Therefore, each apple costs approximately $1.27.

This problem illustrates both the translation of a word problem into a linear equation and the interpretation of the solution in the problem's real-world context.

5.14 Working with Systems of Linear Inequalities

Systems of linear inequalities comprise two or more linear inequalities containing the same set of variables. Like a system of equations, these systems define multiple constraints simultaneously and open up a broad array of valid solutions to the inequalities.

The breadth of valid solutions to these inequalities often gives rise to an area, or a *solution region*, which satisfies all inequalities in the system.

Chapter 5. Inequalities and Systems of Equations

> All feasible solutions to a system of linear inequalities form a solution region on the graph which satisfies all inequalities in the system.

A solution to a system of linear inequalities is a point (or set of points) that makes all the inequalities in the system true. This can be graphically represented as the intersection of all shaded regions of individual inequalities.

To solve a system of linear inequalities, we follow these steps:

1. Solve each inequality individually for y.

2. Graph each inequality on the same set of axes.

3. Identify the region of the coordinate plane, where the regions represented by all inequalities overlap. This region represents the solution to the system.

> Solving a system of linear inequalities involves individually solving each inequality for y, graphing them on the same axes, and finding the overlapping region, which represents the solution.

 Solve the following system of linear inequalities:

$$\begin{cases} y \leq 2x+3 \\ y > -x+1 \end{cases}$$

Solution: To solve this system of linear inequalities, we graph each inequality:

1. Graph $y \leq 2x+3$:

This is a linear equation in slope-intercept form ($y = mx + b$), with a slope of 2 and a y-intercept of 3. The inequality is less than or equal to ($\leq$), so we use a solid line.

2. Graph $y > -x+1$:

This is also in slope-intercept form, with a slope of -1 and a y-intercept of 1. The inequality is a greater than ($>$), so we use a dashed line.

The solution to the system of inequalities is the area of overlap between the two shaded regions:

5.15 Writing Word Problems for Two-Variable Inequalities

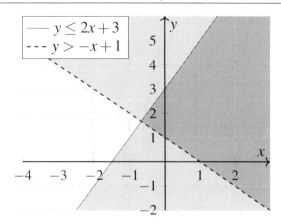

5.15 Writing Word Problems for Two-Variable Inequalities

You will often have a practical scenario in mind and will need to convert that scenario into a system of inequalities.

Key Point

The writing process of a word problem involves a real-world scenario, identifying variables, setting up the system of inequalities, and ensuring the problem is solvable.

Example Consider a local bakery that makes two types of cakes, Chocolate cakes and Red velvet cakes. The costs per cake are $20 and $25 respectively. The bakery has a budget of $500 for the cakes. The counter has space for displaying only 22 cakes at a time. Create a word problem for this scenario.

Solution: This scenario can be translated into the following system of inequalities:

Let x represent the quantity of Chocolate cakes, and y represent the quantity of Red Velvet cakes.

1. The budget constraint: $20x + 25y \leq 500$.

2. The counter space constraint: $x + y \leq 22$.

Solving this system will give the number of each type of cakes that can be made within the constraints.

5.16 Practices

Solve the following inequalities:

1) Solve the inequality $7n > 35$.

2) Solve the inequality $-5x < 10$.

3) Solve the inequality $p + 30 \leq 50$.

4) Solve the inequality $3m - 22 \geq 16$.

5) Solve the inequality $k - 10 > 30$.

Solve:

6) Solve the inequality $5x - 3 > 7$.

7) Solve the inequality $4x + 2 \leq 10$.

8) Solve the inequality $3y - 2 \geq 1$.

9) Solve the inequality $2y + 3 < 7$.

10) Solve the inequality $6z - 4 \geq 2$.

Solve the following compound inequalities:

11) Solve the compound inequality $3 \leq x + 1 < 6$.

12) Solve the compound inequality $-2 + x \geq 0$ and $x - 5 < 1$.

13) Solve the compound inequality $x \div 3 < 2$ or $2x \geq 8$.

14) Solve the compound inequality $-5 \leq x - 3 < -1$.

15) Solve the compound inequality $x - 7 > 3$ and $2x \leq 14$.

Graph the solution region:

16) Graph the inequality $y \leq -2x+4$.

17) Graph the inequality $x > -3$.

18) Graph the inequality $y > x$.

19) Graph the inequality $y \geq -x+3$.

20) Graph the inequality $x \leq 2$.

Graph the solution region:

21) Solve the system $\begin{cases} y \leq 2x+3 \\ y > -x-1 \end{cases}$.

22) Solve the system $\begin{cases} y \geq -3x+2 \\ y \leq x-1 \end{cases}$.

23) Solve the system $\begin{cases} y > 4x-2 \\ y \leq -x+3 \end{cases}$.

24) Solve the system $\begin{cases} y \leq 3x+1 \\ y > -2x-1 \end{cases}$.

25) Solve the system $\begin{cases} y \geq -x+3 \\ y < 2x-1 \end{cases}$.

Solve:

26) Solve $|2x-5| \geq 3$.

27) Solve $|3x+2| \leq 7$.

28) Solve $|4x-1| < 2$.

29) Solve $|5x+3| \geq 8$.

30) Solve $|x-7| \leq 4$.

Solve the system of equations:

31) Solve the following system of equations: $\begin{cases} 2x - 3y = 4 \\ 5x + y = 2 \end{cases}$

32) Solve the following system of equations: $\begin{cases} 4x + 2y = 10 \\ 2x - y = 1 \end{cases}$

33) Solve the following system of equations: $\begin{cases} 3x - y = 4 \\ 2x + 5y = 12 \end{cases}$

34) Solve the following system of equations: $\begin{cases} 5x + 3y = 2 \\ 4x - y = 11 \end{cases}$

35) Solve the following system of equations: $\begin{cases} 7x - 5y = 14 \\ 3x + 2y = 6 \end{cases}$

Solve:

36) Determine the number of solutions in the system of equations: $\begin{cases} 3x + 2y = 12 \\ 6x + 4y = 24 \end{cases}$

37) Determine the number of solutions in the system of equations: $\begin{cases} 10x - 5y = 15 \\ 4x - 2y = 7 \end{cases}$

38) Determine the number of solutions in the system of equations: $\begin{cases} 2x+3y=4 \\ 4x-6y=8 \end{cases}$

39) Determine the number of solutions in the system of equations: $\begin{cases} 7x+y=10 \\ 14x+2y=20 \end{cases}$

40) Determine the number of solutions in the system of equations: $\begin{cases} 5x+2y=15 \\ 10x+4y=35 \end{cases}$

Write a system of equation:

41) Line A: Passes through points $(2,3)$ and $(4,7)$.
Line B: Passes through point $(1,4)$ and has a slope of 2.

42) Line A: Passes through the points $(-1,2)$ and $(1,4)$.
Line B: Passes through point $(-2,-3)$ and has a slope of -1.

43) Line A: Passes through points $(0,0)$ and $(4,-4)$.
Line B: Passes through point $(3,2)$ and has a slope of 1.

44) Line A: Passes through points $(1,1)$ and $(3,3)$.
Line B: Passes through point $(2,-2)$ and has a slope of -3.

45) Line A: Passes through points $(-2,0)$ and $(2,4)$.
Line B: Passes through point $(1,3)$ and has a slope of 0.

Write a system of equations:

46) Anna and Tom go to a book fair. Anna buys 3 adventure novels and 4 mystery novels for $35 while Tom buys 5 adventure novels and 2 mystery novels for $40.

47) Jack and Jill went to the market. Jack bought 6 apples and 2 bananas for $20. Jill bought 4 apples and 5 bananas for $25.

48) A store sells two types of toys. Selling 5 of toy A and 3 of toy B makes $330, while selling 2 of toy A and 4 of toy B makes $240.

49) A baker uses flour and sugar for a cake recipe. 3 kg of flour and 2 kg of sugar costs $13, while 5 kg of flour and 1 kg of sugar costs $15.

50) A movie theater sells adult and child tickets. 4 adult tickets and 3 child tickets cost $50, while 5 adult tickets and 2 child tickets cost $55.

Answer Keys

1) $n > 5$
2) $x > -2$
3) $p \leq 20$
4) $m \geq \frac{38}{3}$
5) $k > 40$
6) $x > 2$
7) $x \leq 2$
8) $y \geq 1$
9) $y < 2$
10) $z \geq 1$
11) $2 \leq x < 5$
12) $2 \leq x < 6$
13) $x < 6$ or $x \geq 4$
14) $-2 \leq x < 2$
15) $\emptyset$
16) See answer details
17) See answer details
18) See answer details
19) See answer details
20) See answer details
21) See answer details
22) See answer details
23) See answer details
24) See answer details
25) See answer details
26) $x \leq 1$ or $x \geq 4$
27) $-3 \leq x \leq \frac{5}{3}$
28) $-\frac{1}{4} < x < \frac{3}{4}$
29) $x \leq -\frac{11}{5}$ or $x \geq 1$
30) $3 \leq x \leq 11$
31) $x = \frac{10}{17}, y = -\frac{16}{17}$
32) $x = \frac{3}{2}, y = 2$
33) $x = \frac{32}{17}, y = \frac{28}{17}$
34) $x = \frac{35}{17}, y = -\frac{47}{17}$
35) $x = 2, y = 0$
36) Infinitely many solutions
37) No solution
38) One unique solution
39) Infinitely many solutions
40) No solution
41) $\begin{cases} y = 2x - 1 \\ y = 2x + 2 \end{cases}$
42) $\begin{cases} y = x + 3 \\ y = -x - 5 \end{cases}$
43) $\begin{cases} y = -x \\ y = x - 1 \end{cases}$
44) $\begin{cases} y = x \\ y = -3x + 4 \end{cases}$
45) $\begin{cases} y = x + 2 \\ y = 3 \end{cases}$
46) $\begin{cases} 3x + 4y = 35 \\ 5x + 2y = 40 \end{cases}$
47) $\begin{cases} 6x + 2y = 20 \\ 4x + 5y = 25 \end{cases}$
48) $\begin{cases} 5x + 3y = 330 \\ 2x + 4y = 240 \end{cases}$
49) $\begin{cases} 3x + 2y = 13 \\ 5x + y = 15 \end{cases}$
50) $\begin{cases} 4x + 3y = 50 \\ 5x + 2y = 55 \end{cases}$

Answers with Explanation

1) Divide both sides by 7 to isolate n.

2) Divide both sides by -5 and flip the inequality sign to isolate x.

3) Subtract 30 from both sides to isolate p.

4) First add 22 to both sides, then divide by 3 to isolate m.

5) Add 10 to both sides to isolate k.

6) To solve this inequality, we start by isolating the variable term by adding 3 to both sides. This gives us $5x > 10$. Then we divide both sides by 5 to solve for x, resulting in $x > 2$. The solution in interval notation is $(2, \infty)$.

7) First subtract 2 from both sides to obtain the inequality $4x \leq 8$. Continue by dividing both sides by 4 to isolate x, which gives $x \leq 2$. The solution in interval notation is $(-\infty, 2]$.

8) First add 2 to both sides to isolate the term with the variable. This simplifies to $3y \geq 3$. Then divide both sides by 3 to isolate y, resulting in $y \geq 1$. The solution in interval notation is $[1, \infty)$.

9) Start by subtracting 3 from both sides of the inequality, which gives $2y < 4$. Then, divide both sides by 2 to isolate y, resulting in $y < 2$. The solution in interval notation is $(-\infty, 2)$.

10) First add 4 to isolate the term with the variable, resulting in $6z \geq 6$. Then divide both sides by 6 to isolate z, resulting in $z \geq 1$. The solution in interval notation is $[1, \infty)$.

11) Subtract 1 from all three parts of the inequality: $3 - 1 \leq x + 1 - 1 < 6 - 1$, then simplify as $2 \leq x < 5$.

12) Simplify each inequality first: $-2 + x \geq 0$ becomes $x \geq 2$ and $x - 5 < 1$ becomes $x < 6$. So the final solution is $2 \leq x < 6$.

13) Solve each inequality: $x \div 3 < 2$ becomes $x < 6$ and $2x \geq 8$ becomes $x \geq 4$. So the solution is $x < 6$ or

$x \geq 4$.

14) Add 3 to all three parts of the inequality, then simplify as $-2 \leq x < 2$.

15) Solve the first inequality: $x - 7 > 3$ becomes $x > 10$. The second inequality becomes $x \leq 7$. The solution has no real numbers as a number cannot be more than 10 and less than or equal to 7 at the same time.

16) For the inequality, the boundary line is $y = -2x + 4$. This line is solid due to "$\leq$". Use (0,0) as your test point. Substituting the coordinates into the inequality, you get $0 \leq -2(0) + 4$ which is true. Therefore, shade the area containing the test point, which is below the line.

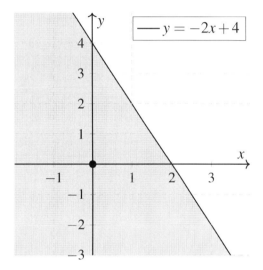

17) The inequality is vertical, so the boundary line is $x = -3$ and it is dashed because of the "$>$" sign. The test point (0,0) leads to $0 > -3$, which is true. So, shade the region to the right of the line.

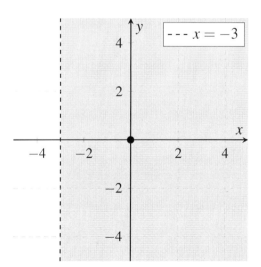

18) The boundary line is $y = x$, which we draw as a dashed line due to the ">" sign. The test point $(0,0)$ is on the line. The test point $(1,0)$ gives $0 > 1$ which is false. Therefore, the shaded region is above the line.

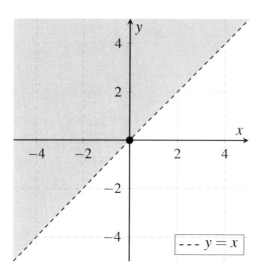

19) The boundary line is $y = -x + 3$, a solid line because of "$\geq$". The test point $(0,0)$ gives $0 \geq -0 + 3$ which is false, thus the shaded region is above the line.

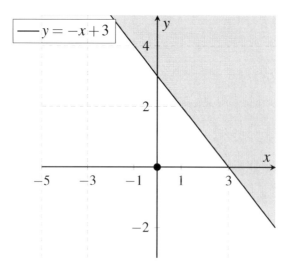

20) This is a vertical line with the equation $x = 2$, a solid line due to the "$\leq$". The test point $(0,0)$ gives $0 \leq 2$ which is true. Hence, we shade the region to the left of the line.

5.16 Practices

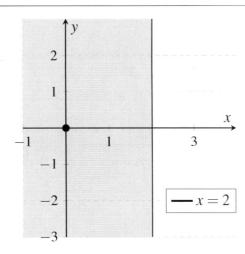

21) After graphing the two inequalities, the common area is below the line $y = 2x + 3$ and above the line $y = -x - 1$. This represents the solution to the inequality system.

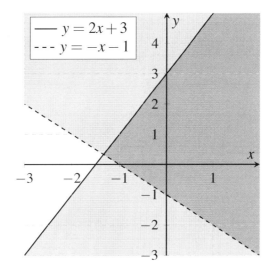

22) After graphing the two inequalities, the common area is above the line $y = -3x + 2$ and below the line $y = x - 1$. This represents the solution to the inequality system.

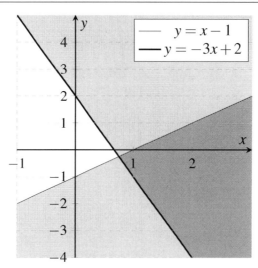

23) After graphing the two inequalities, the common area is above the line $y = 4x - 2$ and below the line $y = -x + 3$. This represents the solution to the inequality system.

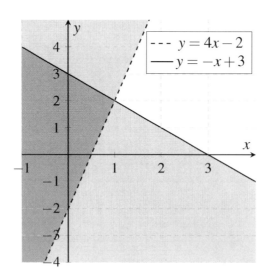

24) After graphing the two inequalities, the common area is below the line $y = 3x + 1$ and above the line $y = -2x - 1$. This represents the solution to the inequality system.

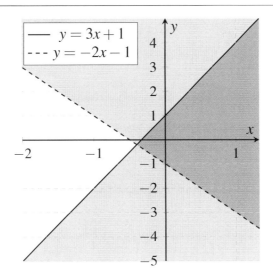

25) After graphing the two inequalities, the common area is above the line $y = -x + 3$ and below the line $y = 2x - 1$. This represents the solution to the inequality system.

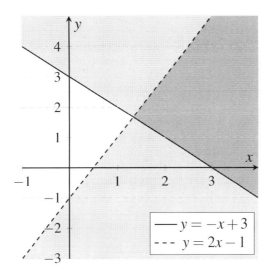

26) Split into $2x - 5 \geq 3$ or $2x - 5 \leq -3$. Solving these gives the solutions $x \leq 1$ or $x \geq 4$.

27) Split into $3x + 2 \leq 7$ and $3x + 2 \geq -7$. Solving these gives the solutions $x \leq \frac{5}{3}$ and $x \geq -3$.

28) Split into $4x - 1 < 2$ and $4x - 1 > -2$. Solving these gives the solutions $-\frac{1}{4} < x < \frac{3}{4}$.

29) Split into $5x + 3 \geq 8$ or $5x + 3 \leq -8$. Solving these gives the solutions $x \leq -\frac{11}{5}$ or $x \geq 1$.

30) Split into $x - 7 \leq 4$ and $x - 7 \geq -4$. Solving these gives the solutions $3 \leq x \leq 11$.

31) We can multiply the second equation by 3 and add it to the first to get $17x = 10$. So $x = \frac{10}{17}$. Substituting $x = \frac{10}{17}$ into the second equation we find $y = 2 - 5 \times \frac{10}{17} = -\frac{16}{17}$.

32) Here, we can solve the second equation for y to get $y = 2x - 1$. Substituting this into the first equation we find $x = \frac{3}{2}$ and $y = 2$.

33) Solve the first equation for y to get $y = 3x - 4$. Substituting this into the second equation we find $x = \frac{32}{17}$ and $y = \frac{28}{17}$.

34) Solve the second equation for y to get $y = 4x - 11$. Substituting this into the first equation we find $x = \frac{35}{17}$ and $y = -\frac{47}{17}$.

35) Here, we can solve the second equation for x to get $x = \frac{6-2y}{3}$. Substituting this into the first equation we get $y = 0$ and substituting $y = 0$ back into the second equation we find $x = 2$.

36) The coefficients and constants ratios are the same ($\frac{3}{6}, \frac{2}{4}$, and $\frac{12}{24}$) which indicates the two equations form the same line hence there are infinitely many solutions.

37) The proportionality of the coefficients is maintained, however, the constants do not match the same ratio ($\frac{10}{4}, \frac{-5}{-2}$, and $\frac{15}{7}$) hence there is no solution as the lines are parallel.

38) The proportions of the coefficients and constants deviate from each other ($\frac{2}{4}, \frac{3}{-6}$, and $\frac{4}{8}$) implying the lines will intersect at one point hence one unique solution.

39) The coefficients and constants ratios are the same ($\frac{7}{14}, \frac{1}{2}$, and $\frac{10}{20}$) which indicates the two equations form the same line hence there are infinitely many solutions.

40) The proportionality of the coefficients is maintained, however, the constants do not match the same ratio ($\frac{5}{10}, \frac{2}{4}$, and $\frac{15}{35}$) hence there is no solution as the lines are parallel.

41) For Line A, the slope $m = \frac{7-3}{4-2} = 2$ and the y-intercept is -1. So, the equation is $y = 2x - 1$. For Line B, we know the slope $m = 2$, and by substituting the point into the equation $y = mx + b$, we get $4 = 2 \times 1 + b$, $b = 2$. So, the equation is $y = 2x + 2$.

42) For Line A, the slope $m = \frac{4-2}{1-(-1)} = 1$ and the y-intercept is 3. So, the equation is $y = x + 3$. For Line B, we know the slope $m = -1$, and by substituting the point into the equation $y = mx + b$, we get $-3 = -1 \times (-2) + b$,

$b = -5$. So, the equation is $y = -x - 5$.

43) For Line A, the slope $m = \frac{-4-0}{4-0} = -1$ and the y-intercept is 0. So, the equation is $y = -x$. For Line B, we know the slope $m = 1$, and by substituting the point into the equation $y = mx + b$, we get $2 = 3 + b$, $b = -1$. So, the equation is $y = x - 1$.

44) For Line A, the slope $m = \frac{3-1}{3-1} = 1$ and the y-intercept is also 0. So, the equation is $y = x$. For Line B, we know the slope $m = -3$, and by substituting the point into the equation $y = mx + b$, we get $-2 = -3 \times 2 + b$, $b = 4$. So, the equation is $y = -3x + 4$.

45) For Line A, the slope $m = \frac{4-0}{2-(-2)} = 1$ and the y-intercept is 2. So, the equation is $y = x + 2$. For Line B, we know the slope $m = 0$. Thus, the line is horizontal, and the y-coordinates for all points on the line is a constant, which is the y-coordinate of the given point. So, the equation is $y = 3$.

46) Let's denote by x the cost of an adventure novel and by y the cost of a mystery novel. Then, we get:
$\begin{cases} 3x + 4y = 35 \\ 5x + 2y = 40 \end{cases}$

47) Let's denote by x the cost of an apple and by y the cost of a banana. Then, we get: $\begin{cases} 6x + 2y = 20 \\ 4x + 5y = 25 \end{cases}$

48) Let's denote by x the cost of toy A and by y the cost of toy B. Then, we get: $\begin{cases} 5x + 3y = 330 \\ 2x + 4y = 240 \end{cases}$

49) Let's denote by x the cost of a kg of flour and by y the cost of a kg of sugar. Then, we get: $\begin{cases} 3x + 2y = 13 \\ 5x + y = 15 \end{cases}$

50) Let's denote by x the cost of an adult ticket and by y the cost of a child ticket. Then, we get:
$\begin{cases} 4x + 3y = 50 \\ 5x + 2y = 55 \end{cases}$

6. Polynomial

6.1 Simplifying Polynomial Expressions

A polynomial is made up of terms, which are either monomials or the sum of monomials. A monomial is an algebraic expression consisting of a single term. Terms that have exactly the same variable part are called "like terms", and we can simplify a polynomial by combining these like terms.

> **Key Point**
>
> When simplifying polynomials, be careful to operate only on terms that have exactly the same variable and power.

We can also simplify polynomial expressions by using the distributive property, also known as the law of distribution, which is a property of real numbers that states that multiplication distributes over addition or subtraction.

> **Key Point**
>
> The distributive property states that $a(b+c) = ab+ac$. This property is also applied when a polynomial is being multiplied by a monomial.

Example Simplify the polynomial expression $7x+5x-3y+2y$.

Solution: First, combine the like terms involving x: $7x+5x = 12x$.

Similarly, combine the like terms involving y: $-3y+2y = -y$.

So, the simplified polynomial is $12x - y$.

 Example Simplify the polynomial expression $2x(3x^2 - 4)$.

Solution: Apply the distributive property: $2x \times 3x^2 = 6x^3$ and $2x \times (-4) = -8x$. So, the simplified polynomial is $6x^3 - 8x$.

6.2 Adding and Subtracting Polynomial Expressions

Adding or subtracting polynomials involves combining like terms which are terms with exactly the same variables raised to the same power. We add or subtract the coefficients of like terms to get the resulting polynomial.

> **Key Point**
>
> When adding or subtracting polynomials, merge only the like terms, the ones that have the same variable with the same power.

Subtracting polynomials is similar, but we have to be very careful with signs. A good strategy is to distribute the "minus" sign to each of the terms of the second polynomial, and then add the polynomials.

> **Key Point**
>
> When subtracting polynomials, first distribute the "minus" sign to each of the terms in the polynomial being subtracted, then proceed with addition of the polynomials.

 Example Add the polynomial expressions $4x^2 + 3x - 2$ and $3x^2 - 5x + 4$.

Solution: Combine like terms to get the sum: $(4x^2 + 3x^2) + (3x - 5x) + (-2 + 4)$, which simplifies to: $7x^2 - 2x + 2$.

 Example Subtract the polynomial expression $3x^2 - 5x + 4$ from $4x^2 + 3x - 2$.

Solution: Distribute the "minus" sign to each term in $-(3x^2 - 5x + 4)$ yielding $-3x^2 + 5x - 4$. Now proceed with addition: $(4x^2 - 3x^2) + (3x + 5x) + (-2 - 4)$, which simplifies to: $x^2 + 8x - 6$.

6.3 Using Algebra Tiles to Add and Subtract Polynomials

Algebra tiles are a visual tool that aid you in understanding the process and rules of adding and subtracting polynomials.

Each tile or set of tiles represents a specific type of term in a polynomial. For instance, a small square represents a constant term, a long rectangular tile represents an x term, and a large square represents an x^2 term. Negative terms are often represented by different coloured tiles.

When adding polynomials using algebra tiles, simply combine the corresponding tiles from each polynomial together. When subtracting, you must introduce the negatives of each term in the second polynomial then combine like tiles.

Key Point

Algebra tiles are a visual tool to help you understand adding and subtracting operations in polynomials. The small square is for constant terms, the rectangle for x terms, and the large square for x^2 terms.

Example Visually add the polynomial expressions $3x^2 + 2x - 1$ and $2x^2 - x + 3$ using algebra tiles.

Solution: We first represent each polynomial with tiles. Then, combine like terms by bringing together the same type of tiles. Counting the combined tiles, the sum is $5x^2 + x + 2$.

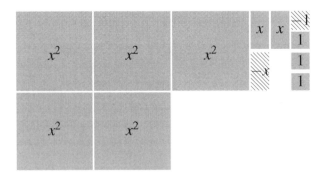

Example Visually subtract the polynomial expression $2x^2 - x + 3$ from $3x^2 + 2x - 1$ using algebra tiles.

Solution: First, we represent each polynomial using tiles. Then, introduce tiles representing the negatives of each term in the polynomial being subtracted. Afterwards, combine like tiles. The resulting polynomial, after counting the remaining tiles, is $x^2 + 3x - 4$.

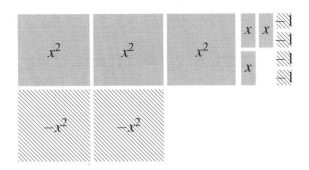

6.4 Multiplying Monomial Expressions

Monomials are algebraic expressions composed of a single term, consisting of just one number, a variable (like x, y, etc.) or a product of numbers and variables.

Multiplication of monomials involves multiplying both coefficients and variables separately. The coefficients are multiplied the normal way, while variables are multiplied using the rules of exponents.

Remember that, when you are multiplying two expressions with the same base, add the exponents. This rule, also known as the Product of Powers Property, states that $a^m \cdot a^n = a^{m+n}$ where a is the common base and m and n are the exponents.

> **Key Point**
>
> Multiplying monomials involves multiplying coefficients and adding exponents of similar variables.

Example Multiply the monomial expressions $4x^2$ and $2x^3$.

Solution: Multiply coefficients 4 and 2 to get $4 \times 2 = 8$.

Multiply variables using the rules of exponents to get $x^{2+3} = x^5$.

Hence the product of the monomial expressions $4x^2$ and $2x^3$ is $8x^5$.

Example Multiply the monomial expressions $3x^2y$ and $2xy^3$.

Solution: Multiply coefficients 3 and 2 to get $3 \times 2 = 6$.

Multiply variables x using the rules of exponents to get $x^{2+1} = x^3$.

Multiply variables y using the rules of exponents to get $y^{1+3} = y^4$.

Hence the product of the monomial expressions $3x^2y$ and $2xy^3$ is $6x^3y^4$.

6.5 Dividing Monomial Expressions

The division of monomials involves working with coefficients and variables separately.

For coefficients, you fragment them as you normally would. For variables, we also use the laws of exponents, but specifically the quotient of powers property, which states that $a^m \div a^n = a^{m-n}$, where a is the base (common for both expressions) and m and n are exponents.

It's crucial to note that division by zero is undefined; so, a monomial with an exponent of zero in the denominator doesn't have a defined value.

Remember, when dividing similar monomials, the result is 1 if and only if the numerators and denominators are identical. This results because of the zero exponent rule which states that any nonzero number raised to the power of 0 is always equal to 1, and this manifests from the quotient of powers property.

🔔 Key Point

Dividing monomials involves dividing coefficients and subtracting exponents of similar variables.

 **Example** Divide the monomials $20x^3y^2$ by $5x^2y$.

Solution: Divide coefficients 20 by 5 to get $20 \div 5 = 4$.

Divide variables x using the rules of exponents to get $x^{3-2} = x$.

Divide variables y using the rules of exponents to get $y^{2-1} = y$.

Hence the quotient of the monomials $20x^3y^2$ and $5x^2y$ is $4xy$.

 **Example** Divide the monomial expressions $6x^3y^2$ by $6x^3y^2$.

Solution: Divide coefficients 6 by 6 to get $6 \div 6 = 1$.

Divide variables x using the rules of exponents to get $x^{3-3} = x^0 = 1$.

Divide variables y using the rules of exponents to get $y^{2-2} = y^0 = 1$.

Multiply resulting values $1 \times 1 \times 1 = 1$

Hence, the quotient of the monomials $6x^3y^2$ and $6x^3y^2$ is 1.

6.6 Multiplying a Polynomial by a Monomial

While a monomial is an algebraic expression consisting of a single term, a polynomial is made up of one or more monomials.

In multiplying a polynomial by a monomial, the monomial multiplies each term of the polynomial separately. This involves multiplying the coefficients and adding the exponents of similarly based variables according to the laws of exponents.

For instance, if we have a monomial a and a polynomial $bx^n + cx^m$, the product would be $abx^n + acx^m$.

The process is described as distribution of the monomial over each term of the polynomial, which is an application of the distributive law.

> Multiplying a polynomial by a monomial involves distributing the monomial to each term of the polynomial.

It is important to note that order does not matter in multiplication, the result will be the same no matter the order in which you multiply the terms.

 Example Multiply the monomial $3xy$ by the polynomial $2x^2y + 5xy^2 + 7$.

Solution: Distribute $3xy$ to each term of the polynomial:

$3xy \times 2x^2y = 6x^3y^2$,

$3xy \times 5xy^2 = 15x^2y^3$,

$3xy \times 7 = 21xy$.

So, the product is $6x^3y^2 + 15x^2y^3 + 21xy$.

 Example Multiply the polynomial $5x^2 + 3x + 7$ by the monomial $4x$.

Solution: Distribute $4x$ to each term of the polynomial:

$4x \times 5x^2 = 20x^3$,

$4x \times 3x = 12x^2$,

$4x \times 7 = 28x$.

So, the product is $20x^3 + 12x^2 + 28x$.

6.7 Using Area Models to Multiply Polynomials

An area model is a visual tool that helps us understand the process of multiplying polynomials.

To multiply two polynomials using an area model, we draw a rectangle and divide it into sections or boxes. Each dimension of the rectangle represents a polynomial and each section represents a term of the resulting polynomial product.

This gives us a more graphical understanding of how each term in one polynomial distributes over each term in the other one, and how the like terms combine.

> An area model visually illustrates the process of multiplying each term in one polynomial with each term in the other and then combining the like terms.

 Example Multiply the polynomial $x+3$ by the polynomial $x+2$ using an area model.

Solution: First, draw a rectangle and divide it into four sections: a square of side x and three rectangles of sides x, 2, and 3.

Label the area of each section with the product of its sides to represent the terms of the product: x^2, $3x$, $2x$, and 6.

Finally, add the areas (or, in terms of algebra, add the terms) to obtain the product of the polynomials: $x^2 + 5x + 6$.

	x	3
x	x^2	$3x$
2	$2x$	6

Example Multiply the polynomials $2x - 1$ and $x + 3$ using an area model.

Solution: Draw a rectangle and divide it into four sections: four rectangles of sides $2x$, x, -1, and 3.

Label the areas of each section with the product of its sides: $2x^2$, $-x$, $6x$, and -3.

Finally, add the areas (or the terms) to obtain the product of the polynomials: $2x^2 + 5x - 3$.

	$2x$	-1
x	$2x^2$	$-x$
3	$6x$	-3

6.8 Multiplying Binomial Expressions

Consider two binomials, $a+b$ and $c+d$. The product of these two binomials is given by the formula:

$$(a+b)(c+d) = ac+ad+bc+bd.$$

This is called the FOIL Method (First, Outer, Inner, Last). First we multiply the first terms in each binomial a and c, then the outer terms a and d, followed by the inner terms b and c, and finally the last terms b and d. Remember to simplify by combining like terms, if any.

> **Key Point**
>
> Multiplication of binomials involves distributing each term in one binomial through each term in the second binomial and combining like terms.

 Find the product of the binomials $x+2$ and $x+3$.

Solution: Applying the FOIL method: $(x+2)(x+3) = x^2+3x+2x+6 = x^2+5x+6$.

 Find the product of the binomials $3y-4$ and $2y+7$.

Solution: Applying the FOIL method: $(3y-4)(2y+7) = 6y^2+21y-8y-28 = 6y^2+13y-28$.

6.9 Using Algebra Tiles to Multiply Binomials

The traditional method of multiplying binomials can often be abstract and difficult to visualize. To help solve this problem, we can use "Algebra Tiles" method. This is a practical way to represent mathematical operations and visualize the distributive property.

Key Point

Algebra tiles are rectangular and square shapes that represent variables and constants. A square tile might represent x^2, a rectangular tile could represent x, and a tiny square tile might represent the constant 1.

Example

Use algebra tiles to multiply the binomials $x+1$ and $x+2$.

Solution: Firstly, represent each term in each binomial with an appropriate tile or set of tiles. For $x+2$, place an x tile and two '1' tiles in a row. Repeat this to form a rectangle. Add the number of each type of tile to find the product:

Number of x^2 tiles $= x \times x = x^2$

Number of x tiles $= x+x+x = 3x$

Number of Constant tiles $= 1+1 = 2$

Therefore, the product of $x+1$ and $x+2$ is x^2+3x+2.

6.10 Factoring Trinomial Expressions

A quadratic trinomial is a polynomial with three terms. It can usually be written in the form ax^2+bx+c, where a, b, and c are constants, and $a \neq 0$.

Factoring trinomial expressions separates them into two binomials. It is a useful technique for simplifying expressions, solving equations, and understanding properties of graphs.

To factor a trinomial, identify two numbers p and q that meet two conditions:

1. Their product equals $\frac{c}{a}$.
2. Their sum equals $\frac{b}{a}$.

You can find these numbers with trial and error or by using the method of factor pairs. The trinomial is then factorized as follows: $ax^2+bx+c = a(x+p)(x+q)$.

Key Point

To factor a trinomial ax^2+bx+c, find two numbers p and q such that their product equals $\frac{c}{a}$ and their sum equals $\frac{b}{a}$. The factorized form is $a(x+p)(x+q)$.

Factorize the trinomial $x^2 + 5x + 6$.

Solution: First, find two numbers p and q such that their product equals 6 and their sum equals 5. The numbers 2 and 3 meet these conditions.

Hence, the factorized form of the trinomial is $(x+2)(x+3)$.

 Factorize the trinomial $2x^2 - 3x - 2$.

Solution: Here, $a = 2$, $b = -3$ and $c = -2$. We need to find two numbers p and q such that their product equals $\frac{c}{a} = \frac{-2}{2} = -1$ and sum is $b = \frac{-3}{2}$.

The numbers -2 and $\frac{1}{2}$ satisfy these conditions. Hence, the factorized form of the trinomial is $2x^2 - 3x - 2 = 2(x-2)(x+\frac{1}{2})$.

6.11 Factoring Polynomial Expressions

Factoring polynomial expressions is vital in simplifying complex expressions and solving equations. When factoring polynomials, our goal is to write the original polynomial as a product of two or more simpler polynomials.

The first step in factoring any polynomial is to find the Greatest Common Factor (GCF). The GCF of a polynomial is the largest polynomial that divides evenly into all terms of the polynomial.

Once the GCF is factored out, we look for special factoring patterns, such as difference of squares, perfect square trinomials, and difference and sum of cubes, to further factorize the remaining polynomial.

> Factoring a polynomial involves expressing it as a product of two or more simpler polynomial expressions. It requires understanding of the distributive property, greatest common factor, and recognizing patterns in polynomials.

 Factorize the polynomial $20x^3 - 10x^2 + 30x$.

Solution: First, determine the GCF of the given polynomial. For $20x^3 - 10x^2 + 30x$, the GCF is $10x$. Factorizing out the GCF, we get: $10x(2x^2 - x + 3)$.

This polynomial cannot be further factorized, so $10x(2x^2 - x + 3)$ is the simplified form of the given polynomial.

 **Example** Factorize the polynomial $x^4 - 81$.

Solution: The given polynomial is a difference of squares, i.e., it is of the form $a^2 - b^2$, which can be factorized as $(a-b)(a+b)$.

Here, $a = x^2$ and $b = 9$. Factoring the polynomial, we get: $x^4 - 81 = (x^2 - 9)(x^2 + 9)$.

The polynomial $x^2 - 9$ is also a difference of squares and can be further factored. Eventually, we get: $x^4 - 81 = (x-3)(x+3)(x^2 + 9)$.

6.12 Using Graphs to Factor Polynomials

Factoring polynomials can often be simplified by using their graphs. The roots of a polynomial are the x-values where the y-value equals to zero. This corresponds to the x-intercepts in the graph of a polynomial. If we figure out these roots from the graph of a polynomial, we can then rewrite the polynomial as a product of multiple polynomials.

Key Point

A factor of a polynomial $p(x)$ is any polynomial $f(x)$ which divides evenly into $p(x)$. If a polynomial $f(x)$ is a factor of $p(x)$, then the root of $f(x)$ will also be the root of $p(x)$. This root will be the x-coordinate at which the graph of $p(x)$ intersects the x-axis.

Note that this method works best with polynomials of degree 2 and degree 3. For polynomials with integer coefficients and roots, you will generally be able to see the integer roots on the graph. Higher degree polynomial functions often have more complex roots, which may not be integers and may extend beyond the parts of the graph that are visible.

Key Point

The Factor Theorem states that the polynomial $p(x) = a_n x^n + a_{n-1} x^{n-1} + \ldots + a_1 x + a_0$ has a factor $x - c$ if $p(c) = 0$.

6.12 Using Graphs to Factor Polynomials

Example Factorize the polynomial $p(x) = x^2 - 7x + 10$.

Solution: Graphing the polynomial $p(x)$, we find that the graph intersects the x-axis at points $x = 2$ and $x = 5$. This means $p(x)$ has roots at $x = 2$ and $x = 5$. According to the Factor Theorem, we can write $p(x)$ as: $p(x) = (x-2)(x-5)$.

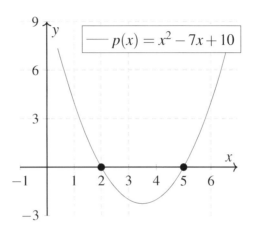

Example Factorize the polynomial $p(x) = x^3 - 8x^2 + 5x + 14$.

Solution: By graphing the polynomial, we find that the graph intersects the x-axis at points $x = -1, x = 2$ and $x = 7$. This means that the roots of the polynomial are $-1, 2$, and 7. Thus, we can express the polynomial $p(x)$ as: $p(x) = (x+1)(x-2)(x-7)$.

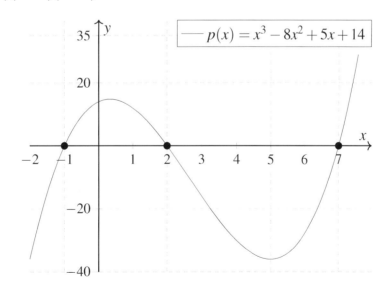

6.13 Factoring Special Case Polynomial Expressions

There are two special cases of polynomial expressions that have unique factoring patterns: perfect square trinomials and difference of squares. Recognizing these patterns will make the factoring process quicker and easier.

Perfect square trinomials refer to polynomial expressions that can be represented as perfect squares of binomials. They follow the pattern $a^2 + 2ab + b^2$, which simplifies to $(a+b)^2$, and $a^2 - 2ab + b^2$, which simplifies to $(a-b)^2$.

Key Point

Perfect square trinomials are polynomials of the form $a^2 + 2ab + b^2 = (a+b)^2$ or $a^2 - 2ab + b^2 = (a-b)^2$.

Example
Factorize the perfect square trinomial $x^2 - 10x + 25$.

Solution: The given polynomial can be expressed as $x^2 - 2 \times 5x + 5^2$, which follows the pattern $a^2 - 2ab + b^2$, where a is x and b is 5. Therefore, the trinomial factored form is $(x-5)^2$.

The **Difference of squares** describe polynomials that can be represented in the form $a^2 - b^2$. This special type of polynomial can be factored using the formula $a^2 - b^2 = (a-b)(a+b)$.

Key Point

The 'difference of squares' is a polynomial of the form $a^2 - b^2$, factored as $(a-b)(a+b)$.

Example
Factorize the difference of squares $x^2 - 9$.

Solution: The given expression can be written as $x^2 - 3^2$, which is an instance of the formula $a^2 - b^2$, where a is x and b is 3. Thus, the factored form of the expression is $(x-3)(x+3)$.

6.14 Using Polynomials to Find Perimeter

When we want to find the perimeter of a geometric figures whose sides are given as polynomials, we add up the polynomial expressions of the lengths of all sides. This gives us a new polynomial which represents the perimeter of the shape.

> **Key Point**
>
> The perimeter of a geometric figure with side lengths given as polynomial expressions can be found by adding all the polynomials.

Example Find the perimeter of a rectangle whose length is $4x+3$ units and width is $2x+5$ units.

Solution: The perimeter P of a rectangle is given by the formula $P = 2l + 2w$, where l is the length and w is the width. Substituting the given expressions, we get $P = 2(4x+3) + 2(2x+5)$. Simplifying, the perimeter is $P = 12x + 16$ units.

Example Find the perimeter of a triangle with sides $3x+2$, $5x+4$, and $2x+3$ units.

Solution: The perimeter P of a triangle is the sum of the lengths of all sides. So, $P = (3x+2) + (5x+4) + (2x+3)$. Simplifying, we find the perimeter to be $P = 10x + 9$ units.

6.15 Practices

Simplify Each Expression:

1) Simplify the polynomial expression $4x^2 - 5y + 2x^2 + 3y$.

2) Simplify the polynomial expression $5y^3 - 2y^3 + 4x - x$.

3) Simplify the polynomial expression $3a^2b - a^2b + 4ab^2 - 2ab^2$.

4) Simplify the polynomial expression $7p^3q^2 - 2p^3q^2 + 5pq^3 - pq^3$.

5) Simplify the polynomial expression $5r^2s - 3r^2s + 4rs^2 - 2rs^2$.

Simplify Each Expression:

6) Simplify the polynomial expression $(3x^2 + 5x - 6) - (2x^2 - x + 3)$.

7) Simplify the polynomial expression $(4x^3 + 3x^2 - x + 2) + (x^3 - 2x^2 + 3x - 6)$.

8) Simplify the polynomial expression $(5x^2 - 2x + 3) - (3x^2 - x - 4)$.

9) Simplify the polynomial expression $(3x^3 - x^2 + 2x - 3) + (2x^3 + 2x^2 - x + 1)$.

10) Simplify the polynomial expression $(2x^3 - 3x^2 + x + 4) - (x^3 - 2x^2 + x - 2)$.

Solve Each Equation:

11) Solve for a: $3a^2b^3 \times 4ab^2 = 72a^2b^5$

12) Solve for x: $5x^2y \times 3x^3y^4 = 75x^4y^5$

13) Solve for m: $2m^3n^2 \times 3mn^3 = 18m^3n^5$

14) Solve for p: $8p^2r^3 \times 2pq^2r^2 = 16p^2q^2r^5$

15) Solve for x: $7x^3z^2 \times 2z^3 = 14x^4z^5$

Evaluate the following monomial expressions:

16) Evaluate $3x^4 \div x^2$ when $x = 2$.

17) Evaluate $10x^7y^3 \div 2x^4y^2$ when $x = 3$ and $y = 2$.

18) Evaluate $6x^3y^2 \div x^2y$ when $x = 2$ and $y = 3$.

19) Evaluate $20x^5y^3 \div 4x^2y^2$ when $x = 4$ and $y = 5$.

20) Evaluate $16x^4y^2 \div 4x^2y$ when $x = 3$ and $y = 2$.

Solve:

21) Multiply the monomial $3x$ by the polynomial $4x^2 + 2x + 7$.

22) Multiply the monomial $2y^2$ by the polynomial $5y^3 + 7y + 8$.

6.15 Practices

23) Multiply the polynomial $6x^2 + 4x + 3$ by the monomial $2x$.

24) Multiply the monomial $4y$ by the polynomial $3y^2 + y + 5$.

25) Multiply the monomial $7x$ by the polynomial $2x^2 + 5x + 6$.

Fill in the Blank:

26) Complete the binomial product using the FOIL method: $(x+3)(x-7) = $ _____ + _____ + _____.

27) Complete the binomial product using the FOIL method: $(2p+5)(3p-1) = $ _____ + _____ + _____.

28) Complete the binomial product using the FOIL method: $(a-5)(2a+3) = $ _____ + _____ + _____.

29) Complete the binomial product using the FOIL method: $(m+3)(4m+5) = $ _____ + _____ + _____.

30) Complete the binomial product using the FOIL method: $(x-2)(y+3) = $ _____ + _____ + _____ + _____.

Solve:

31) Multiply the binomials $x+2$ and $x+3$ using algebra tiles.

32) Multiply the binomials $x+1$ and $x+3$ using algebra tiles.

33) Multiply the binomials $x+2$ and $x+4$ using algebra tiles.

34) Multiply the binomials $x+3$ and $x+4$ using algebra tiles.

35) Multiply the binomials $x+1$ and $x+2$ using algebra tiles.

Factorize the polynomials:

36) $27x^3 - 8$.

37) $x^3 - x^2 - x + 1$.

38) $16x^4 - 81$.

39) $25x^4 - 16$.

40) $x^5 - x^4$.

Solve:

41) Factorize the perfect square trinomial $36y^2 - 12y + 1$.

42) Factorize the difference of squares $81 - 64y^2$.

43) Factorize the perfect square trinomial $p^4 - 14p^2 + 49$.

44) Factorize the difference of squares $16m^4 - 81$.

45) Factorize the perfect square trinomial $a^4 - 4a^2 + 4$.

Solve:

46) The perimeter of a rectangle is $14x + 18$ units. If the length is $4x + 5$ units, what is the width of the rectangle in terms of x?

47) The perimeter of a square is $16x + 20$ units. What is the length of one side of the square?

48) The perimeter of a triangle is $9x + 15$ units. If two sides have lengths $2x + 3$ and $3x + 5$ units, what is the length of the third side?

49) The perimeter of a pentagon is $16x + 25$ units. The lengths of four sides are $3x + 4$, $2x + 3$, $x + 2$, and $x + 3$ units. Find the length of the fifth side.

50) A parallelogram has length $5x + 7$ units and width $3x + 4$ units. What is the perimeter?

Answer Keys

1) $6x^2 - 2y$
2) $3y^3 + 3x$
3) $2a^2b + 2ab^2$
4) $5p^3q^2 + 4pq^3$
5) $2r^2s + 2rs^2$
6) $x^2 + 6x - 9$
7) $5x^3 + x^2 + 2x - 4$
8) $2x^2 - x + 7$
9) $5x^3 + x^2 + x - 2$
10) $x^3 - x^2 + 6$
11) $a = 2$
12) $x = 5$
13) $m = 3$
14) $p = 1$
15) $x = 1$
16) 12
17) 270
18) 36
19) 1600
20) 72
21) $12x^3 + 6x^2 + 21x$
22) $10y^5 + 14y^3 + 16y^2$
23) $12x^3 + 8x^2 + 6x$
24) $12y^3 + 4y^2 + 20y$
25) $14x^3 + 35x^2 + 42x$
26) $x^2, -4x, -21$
27) $6p^2, 13p, -5$
28) $2a^2, -7a, -15$
29) $4m^2, 17m, 15$
30) $xy, 3x, -2y, -6$
31) $x^2 + 5x + 6$
32) $x^2 + 4x + 3$
33) $x^2 + 6x + 8$
34) $x^2 + 7x + 12$
35) $x^2 + 3x + 2$
36) $(3x - 2)(9x^2 + 6x + 4)$
37) $(x - 1)^2(x + 1)$
38) $(4x^2 + 9)(2x + 3)(2x - 3)$
39) $(5x^2 - 4)(5x^2 + 4)$
40) $x^4(x - 1)$
41) $(6y - 1)^2$
42) $(9 + 8y)(9 - 8y)$
43) $(p^2 - 7)^2$
44) $(4m^2 + 9)(2m - 3)(2m + 3)$
45) $(a^2 - 2)^2$
46) $3x + 4$
47) $4x + 5$
48) $4x + 7$
49) $9x + 13$
50) $16x + 22$

Answers with Explanation

1) Combine the like terms involving x^2: $4x^2 + 2x^2 = 6x^2$. Similarly, combine the like terms involving y: $-5y + 3y = -2y$. So, the simplified polynomial is $6x^2 - 2y$.

2) Combine the like terms involving y^3: $5y^3 - 2y^3 = 3y^3$. Similarly, combine the like terms involving x: $4x - x = 3x$. So, the simplified polynomial is $3y^3 + 3x$.

3) Combine the like terms involving a^2b: $3a^2b - a^2b = 2a^2b$. Similarly, combine the like terms involving ab^2: $4ab^2 - 2ab^2 = 2ab^2$. So, the simplified polynomial is $2a^2b + 2ab^2$.

4) Combine the like terms involving p^3q^2: $7p^3q^2 - 2p^3q^2 = 5p^3q^2$. Similarly, combine the like terms involving pq^3: $5pq^3 - pq^3 = 4pq^3$. So, the simplified polynomial is $5p^3q^2 + 4pq^3$.

5) Combine the like terms involving r^2s: $5r^2s - 3r^2s = 2r^2s$. Similarly, combine the like terms involving rs^2: $4rs^2 - 2rs^2 = 2rs^2$. So, the simplified polynomial is $2r^2s + 2rs^2$.

6) Distribute the "minus" sign to all terms on the second polynomial. Next, group and add like terms: $(3x^2 + 5x - 6) - (2x^2 - x + 3) = 3x^2 + 5x - 6 - 2x^2 + x - 3 = x^2 + 6x - 9$.

7) Combine like terms to get the sum: $(4x^3 + 3x^2 - x + 2) + (x^3 - 2x^2 + 3x - 6) = 5x^3 + x^2 + 2x - 4$.

8) Distribute the "minus" sign to all terms on the second polynomial. Next, group and add like terms: $(5x^2 - 2x + 3) - (3x^2 - x - 4) = 5x^2 - 2x + 3 - 3x^2 + x + 4 = 2x^2 - x + 7$.

9) Combine like terms to get the sum: $(3x^3 - x^2 + 2x - 3) + (2x^3 + 2x^2 - x + 1) = 5x^3 + x^2 + x - 2$.

10) Distribute the "minus" sign to all terms on the second polynomial. Next, group and add like terms: $(2x^3 - 3x^2 + x + 4) - (x^3 - 2x^2 + x - 2) = 2x^3 - 3x^2 + x + 4 - x^3 + 2x^2 - x + 2 = x^3 - x^2 + 6$.

11) First, multiply the coefficients and the variables separately on the left-hand side using the product of powers property, which gives $12a^3b^5$. The equation then becomes $12a^3b^5 = 72a^2b^5$. To solve for a, we divide both sides by $12a^2b^5$, which results in $a = 6$.

6.15 Practices

12) First, multiply the coefficients and the variables separately on the left-hand side using the product of powers property, which gives $15x^5y^5$. The equation then becomes $15x^5y^5 = 75x^4y^5$. To solve for x, we divide both sides by $15x^4y^5$, which results in $x = 5$.

13) First, multiply the coefficients and the variables separately on the left-hand side using the product of powers property, which gives $6m^4n^5$. The equation then becomes $6m^4n^5 = 18m^3n^5$. To solve for m, we divide both sides by $6m^3n^5$, which results in $m = 3$.

14) First, multiply the coefficients and the variables separately on the left-hand side using the product of powers property, which gives $16p^3q^2r^5$. The equation then becomes $16p^3q^2r^5 = 16p^2q^2r^5$. To solve for p, we divide both sides by $16p^2q^2r^5$, which results in $p = 1$.

15) First, multiply the coefficients and the variables separately on the left-hand side using the product of powers property, which gives $14x^3z^5$. The equation then becomes $14x^3z^5 = 14x^4z^5$. To solve for x, we divide both sides by $14x^3z^5$, which results in $x = 1$.

16) First simplify the expression by using the rules of exponents, which yields $3x^{4-2} = 3x^2$. Then substitute $x = 2$ into the simplified expression, getting $3 \times 2^2 = 12$.

17) First simplify the expression by using the rules of exponents, which leads us to $5x^3y = 5 \times 3^3 \times 2 = 270$.

18) First simplify the expression by using the rules of exponents, which leads us to $6xy = 6 \times 2 \times 3 = 36$.

19) First simplify the expression by using the rules of exponents, which leads to $5x^3y = 5 \times 4^3 \times 5 = 1600$.

20) First simplify the expression by using the rules of exponents, which leads to $4x^2y = 4 \times 3^2 \times 2 = 72$.

21) This is done by distributing $3x$ to each term of the polynomial: $3x \times 4x^2 = 12x^3$ and $3x \times 2x = 6x^2$ and $3x \times 7 = 21x$.

22) This is done by distributing $2y^2$ to each term of the polynomial: $2y^2 \times 5y^3 = 10y^5$ and $2y^2 \times 7y = 14y^3$ and $2y^2 \times 8 = 16y^2$.

23) This is done by distributing $2x$ to each term of the polynomial: $2x \times 6x^2 = 12x^3$ and $2x \times 4x = 8x^2$ and $2x \times 3 = 6x$.

24) This is done by distributing $4y$ to each term of the polynomial: $4y \times 3y^2 = 12y^3$ and $4y \times y = 4y^2$ and $4y \times 5 = 20y$.

25) This is done by distributing $7x$ to each term of the polynomial: $7x \times 2x^2 = 14x^3$ and $7x \times 5x = 35x^2$ and $7x \times 6 = 42x$.

26) Using the FOIL method, the result is $x^2 - 7x + 3x - 21$. This simplifies to $x^2 - 4x - 21$.

27) Applying the FOIL method, the result is $6p^2 - 2p + 15p - 5$. This simplifies to $6p^2 + 13p - 5$.

28) Use the FOIL method: $2a^2 + 3a - 10a - 15$. Simplified, this gives $2a^2 - 7a - 15$.

29) By using the FOIL method we have $4m^2 + 12m + 5m + 15$. Simplified, this equals $4m^2 + 17m + 15$.

30) The product from using the FOIL method is $xy + 3x - 2y - 6$.

31) For $x+2$, place an 'x' tile and two '1' tiles, and for $x+3$, place an 'x' tile and three '1' tiles. Form a rectangle and count each tile:

Number of x^2 tiles $= x \times x = x^2$.

Number of x tiles $= 2x + 3x = 5x$.

Number of constant tiles $= 1 + 1 + 1 + 1 + 1 + 1 = 6$.

Therefore, the product of $x+2$ and $x+3$ is $x^2 + 5x + 6$.

32) For $x+1$, place an 'x' tile and a '1' tile, and for $x+3$, place an 'x' tile and three '1' tiles. Form a rectangle and count each tile:

Number of x^2 tiles $= x \times x = x^2$.

Number of x tiles $= x + 3x = 4x$.

Number of constant tiles $= 1 + 1 + 1 = 3$.

Therefore, the product of $x+1$ and $x+3$ is $x^2 + 4x + 3$.

33) For $x+2$, place an 'x' tile and two '1' tiles, and for $x+4$, place an 'x' tile and four '1' tiles. Form a rectangle and count each tile:

Number of x^2 tiles $= x \times x = x^2$.

Number of x tiles $= 2x + 4x = 6x$.

Number of constant tiles $= 1 + 1 + 1 + 1 + 1 + 1 + 1 + 1 = 8$.

6.15 Practices

Therefore, the product of $x+2$ and $x+4$ is x^2+6x+8.

34) For $x+3$, place an 'x' tile and three '1' tiles, and for $x+4$, place an 'x' tile and four '1' tiles. Form a rectangle and count each tile:

Number of x^2 tiles $= x \times x = x^2$.

Number of x tiles $= 3x + 4x = 7x$.

Number of constant tiles $= 1+1+1+1+1+1+1+1+1+1+1+1 = 12$.

Therefore, the product of $x+3$ and $x+4$ is $x^2+7x+12$.

35) For $x+1$, place an 'x' tile and a '1' tile, and for $x+2$, place an 'x' tile and two '1' tiles. Form a rectangle and count each tile:

Number of x^2 tiles $= x \times x = x^2$.

Number of x tiles $= x + 2x = 3x$.

Number of constant tiles $= 1 + 1 = 2$.

Therefore, the product of $x+1$ and $x+2$ is x^2+3x+2.

36) The expression is a difference of cubes which can be factorized into $(a-b)(a^2+ab+b^2)$. Here a is $3x$ and b is 2.

37) The expression can be rearranged as $x^3 - x - x^2 + 1$, which can be factored into $(x^2 - 2x + 1)(x + 1)$, simplifying further to $(x-1)^2(x+1)$.

38) The given expression is a difference of squares, and can be factored as $(4x^2+9)(4x^2-9)$, and then $(4x^2+9)(2x+3)(2x-3)$.

39) The given expression is a difference of squares and can be factored into $(a-b)(a+b)$, where a is $5x^2$ and b is 4.

40) The GCF of the expression is x^4. When we factorize out x^4, we get $x^4(x-1)$.

41) It can be written as $36y^2 - 2 \times 6y \times 1 + 1^2$, which is the form $a^2 - 2ab + b^2$ where a is $6y$ and b is 1. So, the factored form is $(6y-1)^2$.

42) It can be written as $9^2 - (8y)^2$, which can be factored as $(9+8y)(9-8y)$.

43) It can be written as $p^4 - 2 \times 7 \times p^2 + 7^2$, hence it is a perfect square trinomial in the form $a^2 - 2ab + b^2$, so the factored form is $(p^2 - 7)^2$.

44) It can be written as $(4m^2)^2 - 9^2$, which can be factored as $(4m^2 + 9)(4m^2 - 9) = (4m^2 + 9)(2m - 3)(2m + 3)$.

45) It can be written as $a^4 - 2 \times 2 \times a^2 + 2^2$, hence it is a perfect square trinomial in the form $a^2 - 2ab + b^2$, so the factored form is $(a^2 - 2)^2$.

46) We use the formula for the perimeter of a rectangle: $P = 2l + 2w$. Rearranging for w and substituting $P = 14x + 18$ and $l = 4x + 5$, we get $w = \frac{P}{2} - l$. Substituting and simplifying, we find $w = 3x + 4$ units.

47) Since a square has all sides equal, the side length is $\frac{P}{4} = \frac{16x+20}{4}$, which simplifies to $4x + 5$ units.

48) Calling the third side s, we can write $P = s + (2x + 3) + (3x + 5)$. Rearranging for s and substituting $P = 9x + 15$, we find $s = 4x + 7$ units.

49) Let s represent the length of the fifth side. We know that $P = s + (3x + 4) + (2x + 3) + (x + 2) + (x + 3)$. Solving for s and substituting $P = 16x + 25$, we find $s = 9x + 13$ units.

50) The perimeter P of a parallelogram is given by the formula $P = 2l + 2w$, where l is the length and w is the width. For the given length $5x + 7$ and width $3x + 4$, the perimeter can be calculated as $P = 2(5x + 7) + 2(3x + 4) = 10x + 14 + 6x + 8 = 16x + 22$ units.

7. Quadratic

7.1 Solving Quadratic Equations

A quadratic equation is an equation of the second degree, meaning it contains at least one term that is squared. The standard form of a quadratic equation is $ax^2 + bx + c = 0$, where a, b, and c are constants and x represents an unknown variable.

> **Key Point**
>
> A quadratic equation $ax^2 + bx + c = 0$ can have either two distinct solutions, one repeated solution, or no real solution, depending on the value of the discriminant $(b^2 - 4ac)$.

If the discriminant is greater than 0, then the quadratic equation has two distinct real roots. If the discriminant equals 0, then the equation has exactly one real root (also called a repeated root). If the discriminant is less than 0, then the equation has no real roots, but it has two distinct complex roots.

The solutions to a quadratic equation are given by the quadratic formula:

$$x = \frac{-b \pm \sqrt{b^2 - 4ac}}{2a}.$$

> **Key Point**
>
> The discriminant $(b^2 - 4ac)$ determines a quadratic equation's roots: positive, for two distinct real roots, zero, for one repeated real root, negative, for two complex roots. Roots are given by $x = \frac{-b \pm \sqrt{b^2 - 4ac}}{2a}$.

Example Solve the quadratic equation $x^2 - 3x - 4 = 0$.

Solution: For the given equation $x^2 - 3x - 4 = 0$, we have $a = 1$, $b = -3$, and $c = -4$.

Applying the quadratic formula:

$$x = \frac{-(-3) \pm \sqrt{(-3)^2 - 4(1)(-4)}}{2(1)} \Rightarrow x = \frac{3 \pm \sqrt{25}}{2} = \frac{3 \pm 5}{2}.$$

So, the two solutions are $x = 4$ and $x = -1$.

 Example Solve the quadratic equation $x^2 - 2x + 1 = 0$.

Solution: For the equation $x^2 - 2x + 1 = 0$, we have $a = 1$, $b = -2$, and $c = 1$.

Applying the quadratic formula, we get:

$$x = \frac{-(-2) \pm \sqrt{(-2)^2 - 4(1)(1)}}{2(1)} = \frac{2 \pm \sqrt{4 - 4}}{2} = \frac{2}{2} = 1.$$

Since the discriminant is zero, this equation has one real solutions.

Example Solve the quadratic equation $2x^2 - 2x + 1 = 0$.

Solution: For the equation $2x^2 - 2x + 1 = 0$, we have $a = 2$, $b = -2$, and $c = 1$.

Applying the quadratic formula, we get:

$$x = \frac{-(-2) \pm \sqrt{(-2)^2 - 4(2)(1)}}{2(2)} = \frac{2 \pm \sqrt{4 - 8}}{4} = \frac{2 \pm \sqrt{-4}}{4}.$$

Since the square root of a negative number is not a real number, this equation has no real solutions.

7.2 Graphing Quadratic Functions

The standard form of a quadratic function is $y = ax^2 + bx + c$. The graph of a quadratic function is a curve called a parabola, which can either open upwards or downwards.

7.2 Graphing Quadratic Functions

🔔 Key Point

In the quadratic function $y = ax^2 + bx + c$, the parabola always opens upwards when $a > 0$ and downwards when $a < 0$.

Quadratic functions have a specific structure and standard features.

🔔 Key Point

The highest or the lowest point of a parabola is called the vertex. The vertex of a quadratic function $f(x) = ax^2 + bx + c$ is given by the formula (h, k), where $h = \frac{-b}{2a}$ and $k = f(h)$.

🔔 Key Point

The line of symmetry of a parabola passes through the vertex and is represented by the equation $x = h$.

🔔 Key Point

The y-intercept of a function is the point where the graph crosses the y-axis. This point is always $(0, c)$, where c is the constant term in the equation of the function.

🔔 Key Point

The x-intercept(s) of a function are the points where the graph crosses the x-axis. To find these points, we set $y = 0$ in the equation and solve for x.

📋 Example

Graph the quadratic function $f(x) = x^2 - 4x + 3$.

Solution: First, we identify $a = 1$, $b = -4$, and $c = 3$ from the function.

Next, we calculate the vertex. The h-value is given by the formula $\frac{-b}{2a} = \frac{-(-4)}{2 \times 1} = 2$.

Substituting $x = 2$ into the function gives us the k-value: $f(2) = (2)^2 - 4(2) + 3 = -1$. So, the vertex is at $(2, -1)$.

The line of symmetry passes through the vertex and is $x = 2$.

The y-intercept is found by setting $x = 0$: $f(0) = (0)^2 - 4(0) + 3 = 3$. So, the y-intercept is at $(0, 3)$.

The x-intercept(s) are found when $y = 0$, or $0 = x^2 - 4x + 3$. Solving this quadratic equation, we find the roots are 1 and 3, so the x-intercepts are $(1, 0)$ and $(3, 0)$.

Given these key features, we can graph the quadratic function as follow:

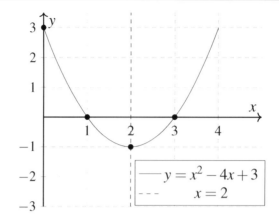

7.3 Factoring to Solve Quadratic Equations

Factoring is one of the methods used to solve quadratic equations. A quadratic equation, given in the form $ax^2 + bx + c = 0$, can be solved by breaking it down, or factoring it, into two binomial equations.

To factor a quadratic equation, one must find two numbers that add up to b and multiply to c. These two numbers are used to break down or factor the bx term, which results in a quadratic equation that is easier to solve. This method is only applicable when the quadratic is factorable and the lead coefficient $a = 1$.

After factoring a quadratic equation, we can use the multiplication property of zero, which states that if $p \times q = 0$, then either $p = 0$ or $q = 0$.

🔔 Key Point

To solve a quadratic equation $x^2 + bx + c = 0$ by factoring, find two numbers that add to b and multiply to c. Use these numbers to factor the equation, then apply the zero product property ($p \times q = 0$ implies $p = 0$ or $q = 0$) to find the solutions.

📋 Example

Factor and solve the quadratic equation $x^2 + 5x + 6 = 0$.

Solution: In this equation, $a = 1, b = 5, c = 6$. We must find two numbers that add up to 5 (the b value) and multiply to 6 (the c value). The numbers 2 and 3 satisfy this. So, we have: $x^2 + 5x + 6 = (x+2)(x+3) = 0$.

To find the solution, we set each factor equal to zero and solve for x, this gives: $x + 2 = 0 \Rightarrow x = -2$ and $x + 3 = 0 \Rightarrow x = -3$.

So, the solutions to the equation $x^2 + 5x + 6 = 0$ are $x = -2$ and $x = -3$.

7.4 Understanding Transformations of Quadratic Functions

The vertex form of a quadratic function $f(x) = ax^2 + bx + c$ is $f(x) = a(x-h)^2 + k$.

These functions are characterized by their parabolic graphs. The graph of the function might be shifted, flipped, or stretched, depending on the values of a, h, and k. These changes are called transformations.

Key Point

If $f(x) = a(x-h)^2 + k$, the coefficient a determines the **vertical stretch or compression** and direction (upwards or downwards) of a parabola. The value of h is related to the **horizontal shift** of the parabola (right or left), and k represents the **vertical shift** (up or down).

- if $|a| > 1$, the parabolic graph is stretched vertically,
- if $|a| < 1$, the parabolic graph is compressed vertically,
- if $h > 0$, the parabolic graph shifts h units to the right,
- if $h < 0$, the parabolic graph shifts h units to the left,
- if $k > 0$, the parabolic graph shifts up by k units, and
- if $k < 0$, the parabolic graph shifts downwards by k units.

Example Determine the transformations of the quadratic function $f(x) = -2x^2 + 4x - 3$.

Solution: The vertex form of $f(x) = -2x^2 + 4x - 3$ is $f(x) = -2(x-1)^2 - 1$

1. Since $a = -2 < 0$, the parabola opens downwards. Also, the absolute value of a is 2, which indicates there is a vertical stretch by a factor of 2.

2. We have $h = 1 > 0$. Thus, the graph is horizontally shifted 1 unit towards the right.

3. Since $k = -1 < 0$, there is a vertical shift 1 units downward.

Hence, the function $f(x) = -2x^2 + 4x - 3$ has been vertically stretched by a factor of 2, reflected over the x-axis, shifted to the right by 1 unit, and then shifted downward by 1 units.

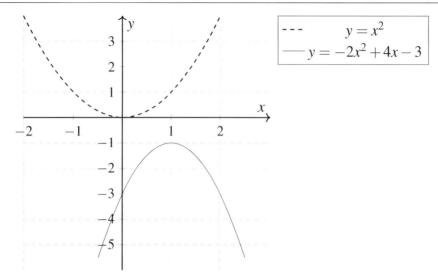

7.5 Completing Function Tables for Quadratic Functions

The function table contains values of x and according values of y, where y is computed using the formula of the quadratic function. Completing the function table for a quadratic function helps in understanding the relationship between the variables and provides a concrete method for graphing the function.

Key Point

Function tables for quadratic functions are completed by substituting the x values into the quadratic equation to find the corresponding y values.

Example

Given the quadratic function $f(x) = x^2 + 2x + 1$, complete the following function table:

x	-2	-1	0	1	2
$f(x)$					

Solution: The function table is completed by substituting the given x values into the function.

For $x = -2$, $f(-2) = (-2)^2 + 2(-2) + 1 = 4 - 4 + 1 = 1$.

For $x = -1$, $f(-1) = (-1)^2 + 2(-1) + 1 = 1 - 2 + 1 = 0$.

For $x = 0$, $f(0) = 0 + 0 + 1 = 1$.

For $x = 1$, $f(1) = 1 + 2 + 1 = 4$.

For $x = 2$, $f(2) = 4 + 4 + 1 = 9$.

So the completed function table is:

x	-2	-1	0	1	2
$f(x)$	1	0	1	4	9

7.6 Determining Domain and Range of Quadratic Functions

The domain of a function encompasses all the possible input values (often denoted as x), whereas the range includes all the possible output values (denoted as y or $f(x)$).

> **Key Point**
>
> The domain of a quadratic functions $f(x) = ax^2 + bx + c$ is all real numbers, $x \in \mathbb{R}$.

For the range, it depends on whether the graph opens upwards, which occurs when $a > 0$ or downwards when $a < 0$.

> **Key Point**
>
> In the quadratic function $f(x) = ax^2 + bx + c$, if $a > 0$, the range is $[y_{vertex}, +\infty)$. On the other hand, if $a < 0$, the range is $(-\infty, y_{vertex}]$.

Example Determine the domain and range of the quadratic function $f(x) = 2x^2 + 4x + 1$.

Solution: The domain of any quadratic function is all real numbers, i.e., $x \in \mathbb{R}$.

To evaluate the range, we first need to determine the vertex of the function.

The x-coordinate of the vertex is given by $-\frac{b}{2a} = -\frac{4}{4} = -1$. Substituting this value back to the function, we find the y-coordinate of the vertex, $f(-1) = 2(-1)^2 + 4(-1) + 1 = -1$.

Since $a = 2$ is positive, the graph of the function opens upwards. Therefore, the range of this function is $[-1, +\infty)$.

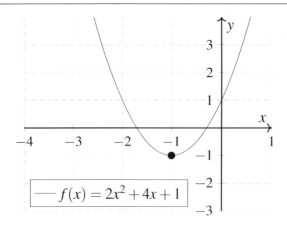

Example Determine the domain and range of the quadratic function $f(x) = -3x^2 + 6x - 1$.

Solution: The domain of any quadratic function is all real numbers, thus, $x \in \mathbb{R}$.

For the range, we need to find the vertex of the function.

The x-coordinate of the vertex is given by $-\frac{b}{2a} = -\frac{6}{-6} = 1$. Substituting this back to the function yields the y-coordinate of the vertex, $f(1) = -3(1)^2 + 6(1) - 1 = 2$.

Since $a = -3$ is negative, the graph opens downwards. Hence, the range of this function is $(-\infty, 2]$.

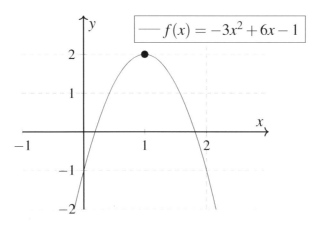

7.7 Factoring Special Case Quadratics

Factoring special case quadratics involve recognizing certain patterns in quadratic expressions to break them down into simpler, solvable components. Two significant cases of special quadratics include the perfect square trinomials and the difference of squares.

A perfect square trinomial is a trinomial that can be factored into the square of a binomial. It is a special type of $(ax)^2 + 2abx + b^2$ and can be factored to $(ax + b)^2$.

7.8 Using Algebra Tiles to Factor Quadratics

The factoring formula for a perfect square trinomial $(ax)^2 + 2abx + b^2$ is $(ax+b)^2$.

On the other hand, the difference of squares is a binomial which follows the pattern $a^2 - b^2$. It can be factored into $(a+b)(a-b)$.

The factoring formula for the difference of squares $a^2 - b^2$ is $(a+b)(a-b)$.

Example Factor the quadratic trinomial $x^2 + 10x + 25$.

Solution: Recognizing the perfect square trinomial structure, we can break down $x^2 + 10x + 25$ using the formula $(ax+b)^2$, where both $a = 1$ and $b = 5$. Therefore, $x^2 + 10x + 25$ factors to $(x+5)^2$.

Example Factor the quadratic binomial $16x^2 - 81$.

Solution: Observing the structure of the difference of squares, we can break down $16x^2 - 81$ using the formula $(a+b)(a-b)$, where $a = 4x$ and $b = 9$. Therefore, $16x^2 - 81$ factors to $(4x+9)(4x-9)$.

7.8 Using Algebra Tiles to Factor Quadratics

Algebra tiles are square and rectangle-shaped tiles that represent an area model of algebraic expressions and can be used for operations like factoring.

For quadratic expressions, a square tile is used to represent x^2, a rectangular tile to represent x, and a small square tile to signify 1. Laying out these tiles to form a perfect square helps us visualize the factors of a quadratic equation.

Algebra tiles visually represent algebraic expressions. For quadratics, x^2, x, and 1 are respectively represented by square, rectangular, and small square tiles.

Example Factor the quadratic expression $x^2 + 7x + 10$ using algebra tiles.

Solution: Using algebra tiles, the expression $x^2 + 7x + 10$ can be represented as one x^2 tile, seven x tiles and ten 1 tiles.

We arrange these tiles to form a perfect square. From the arrangement, we observe that the factors are represented by the lengths of the sides of the square, which equate to $x+2$ and $x+5$. So, the factors of $x^2 + 7x + 10$ are $(x+2)$ and $(x+5)$.

7.9 Writing Quadratic Functions from Vertices and Points

The vertex form of a quadratic function is given by $f(x) = a(x-h)^2 + k$ where (h,k) is the vertex of the parabola and a is a non-zero constant. Given the vertex and a point on the graph of the function, we can solve for the value of a.

Key Point

$f(x) = a(x-h)^2 + k$ is the vertex form of a quadratic function where (h,k) is the vertex and a is a non-zero constant. The vertex form can be written using given vertex and a point on the function.

Example

A baseball is hit and its height h in feet after t seconds is given by a quadratic function. The vertex of the parabola is at $(2,30)$, and the ball hits the ground after 5 seconds. Find the quadratic function that models this situation.

Solution: Firstly, the fact that the ball hits the ground in 5 seconds gives us another point on the graph $(5,0)$.

Next, we setup the vertex form of the quadratic function using the vertex $(h,k) = (2,30)$: $h(t) = a(t-2)^2 + 30$.

To find a, we substitute the coordinates of the point $(5,0)$ into the equation: $0 = a(5-2)^2 + 30$. This simplifies to: $0 = 9a + 30$. Solving for a gives us $a = -\frac{10}{3}$.

The quadratic function which models the height of the baseball over time is: $h(t) = -\frac{10}{3}(t-2)^2 + 30$.

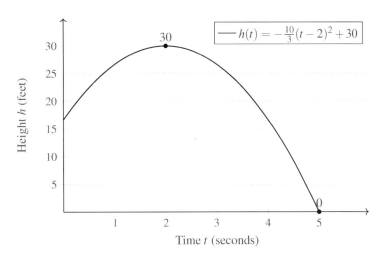

7.10 Practices

Solve:

1) Solve the quadratic equation $3x^2 - 5x + 2 = 0$.

2) Solve the quadratic equation $x^2 - 4x + 3 = 0$.

3) Solve the quadratic equation $2x^2 + 3x + 1 = 0$.

4) Solve the quadratic equation $x^2 - 2x - 3 = 0$.

5) Solve the quadratic equation $4x^2 - 12x + 9 = 0$.

Graphical Questions:

6) Sketch the graph of the quadratic function $y = x^2 - 2x + 3$.

7) Given the graph of the function $f(x) = ax^2 + bx + c$, identify the function's vertex, y-intercept, and x-intercepts.

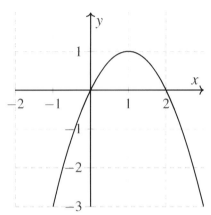

8) Draw the graph of the quadratic function $f(x) = 2x^2 + 4x + 1$.

9) Based on the graph, determine the orientation, vertex, y-intercept, and x-intercepts (if any) of the quadratic function $y = ax^2 + bx + c$.

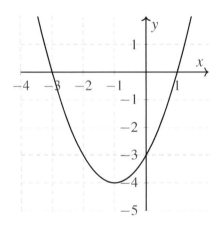

10) Sketch and label the graph of the quadratic function $y = -3x^2 + 6x + 5$.

Fill in the blanks:

11) Fill in the blanks to factorize the quadratic equation $x^2 + _____ x + 10 = 0$ such that the numbers add up to 7 and multiply to 10.

12) Fill in the blanks in the equation $(x + _____)(x + _____) = 0$ to find the roots of the equation $x^2 + 9x + 18 = 0$.

7.10 Practices

13) Complete the factoring of the quadratic equation $x^2 + 6x + 9 = 0$ using the formula $(x + \underline{\hspace{2cm}})^2$.

14) If the quadratic equation is in the form of $x^2 + bx + c = 0$, the two numbers that factorize this expression must add up to \underline{\hspace{1.5cm}} and multiply to \underline{\hspace{1.5cm}}.

15) Write the solutions of the equation $x^2 + 4x + 4 = 0$ in the format $x = \underline{\hspace{2cm}}, \underline{\hspace{2cm}}$.

Solve:

16) Given the quadratic function $f(x) = x^2 + 4x + 4$, solve for $f(0)$, $f(1)$ and $f(-1)$.

Solve:

17) Solve for the domain and range of the quadratic function $f(x) = x^2 - 4x + 4$.

18) Solve for the domain and range of the quadratic function $f(x) = -2x^2 + 6x - 3$.

19) Solve for the domain and range of the quadratic function $f(x) = 3x^2 - 12x + 7$.

20) Solve for the domain and range of the quadratic function $f(x) = -x^2 + 4x - 2$.

21) Solve for the domain and range of the quadratic function $f(x) = 4x^2 - 16x + 9$.

Fill in the Blank:

22) Factorize the expression $x^2 + 8x + 15 = \underline{\hspace{2cm}}$.

23) Factorize the expression $x^2 + 10x + 21 = \underline{\hspace{2cm}}$.

24) Factorize the expression $x^2 + 5x + 6 = \underline{\hspace{2cm}}$.

25) Factorize the expression $x^2 + 14x + 49 = \underline{\hspace{2cm}}$.

26) Factorize the expression $x^2 + 4x + 4 = \underline{\hspace{2cm}}$.

Solve:

27) Given that the vertex of a quadratic function is at $(2,-3)$, and the function passes through the point $(4,1)$, find the equation of the quadratic function.

28) Given that a quadratic function with vertex at $(-1,5)$ passes through the point $(2,-4)$, what is the equation of the quadratic function?

29) A basketball player throws a ball. The vertex of the parabolic path of the ball is at $(2,10)$, and the ball hits the ground after 4 seconds. Write the quadratic function in vertex form modeling this situation.

30) If the vertex of a quadratic function is $(-1,2)$ and it passes through the point $(1,6)$, what is the equation of the quadratic function?

31) A stone is thrown and its height h in feet after t seconds is given by a quadratic function. The vertex of the parabola is at $(1,4)$, and the stone hits the ground after 3 seconds. Find the quadratic function that models this situation.

Problem Solving:

32) Find the quadratic function that opens upwards, has a y-intercept of -3, and its vertex is at $(1,-4)$.

33) Find the quadratic function with vertex at $(2,3)$, opening upwards, and passing through the point $(1,4)$.

34) Find the equation of a function with vertex at the origin $(0,0)$, opens downwards, and passing through the point $(2,-4)$.

35) Find the quadratic function that opens downwards, has a vertex at $(1,2)$, and a y-intercept at 1.

36) Create a quadratic function that opens downwards, has a vertex at $(4,2)$, and passes through point $(-2,-10)$.

Answer Keys

1) $x = 1$ and $x = \frac{2}{3}$

2) $x = 1$ and $x = 3$

3) $x = -1$ and $x = -\frac{1}{2}$

4) $x = -1$ and $x = 3$

5) $x = \frac{3}{2}$ (repeated solution)

6) See answer details

7) $(1,1), (0,0), (0,0)$ and $(2,0)$

8) See answer details

9) Upwards, $(-1,-4), (0,-3), (-3,0)$ and $(1,0)$

10) See answer details

11) 2, 5

12) 3, 6

13) 3

14) b, c

15) $-2, -2$

16) $f(0) = 4, f(1) = 9, f(-1) = 1.$

17) Domain: $x \in \mathbb{R}$, Range: $[0, +\infty)$

18) Domain: $x \in \mathbb{R}$, Range: $(-\infty, 1.5]$

19) Domain: $x \in \mathbb{R}$, Range: $[-5, +\infty)$

20) Domain: $x \in \mathbb{R}$, Range: $(-\infty, 2]$

21) Domain: $x \in \mathbb{R}$, Range: $[-7, +\infty)$

22) $(x+3)(x+5)$

23) $(x+7)(x+3)$

24) $(x+2)(x+3)$

25) $(x+7)^2$

26) $(x+2)^2$

27) $f(x) = (x-2)^2 - 3$

28) $f(x) = -(x+1)^2 + 5$

29) $f(t) = -\frac{5}{2}(t-2)^2 + 10$

30) $f(x) = (x+1)^2 + 2$

31) $h(t) = -(t-1)^2 + 4$

32) $f(x) = (x-1)^2 - 4$

33) $f(x) = (x-2)^2 + 3$

34) $f(x) = -x^2$

35) $f(x) = -(x-1)^2 + 2$

36) $f(x) = -\frac{1}{3}(x-4)^2 + 2$

Answers with Explanation

1) For the given equation $3x^2 - 5x + 2 = 0$, $a = 3$, $b = -5$, and $c = 2$. Applying the quadratic formula: $x = \frac{-(-5) \pm \sqrt{(-5)^2 - 4 \times 3 \times 2}}{2 \times 3} = \frac{5 \pm \sqrt{25-24}}{6} = \frac{5 \pm 1}{6}$. So, the solutions are $x = 1$ and $x = \frac{2}{3}$.

2) For the given equation $x^2 - 4x + 3 = 0$, $a = 1$, $b = -4$, and $c = 3$. Applying the quadratic formula: $x = \frac{-(-4) \pm \sqrt{(-4)^2 - 4 \times 1 \times 3}}{2 \times 1} = \frac{4 \pm \sqrt{16-12}}{2} = \frac{4 \pm 2}{2}$. So, the solutions are $x = 1$ and $x = 3$.

3) For the equation $2x^2 + 3x + 1 = 0$, $a = 2$, $b = 3$, and $c = 1$. Applying the quadratic formula: $x = \frac{-3 \pm \sqrt{3^2 - 4 \times 2 \times 1}}{2 \times 2} = \frac{-3 \pm \sqrt{9-8}}{4} = \frac{-3 \pm 1}{4}$. So, the solutions are $x = -1$ and $x = -\frac{1}{2}$.

4) For the given equation $x^2 - 2x - 3 = 0$, $a = 1$, $b = -2$, and $c = -3$. Applying the quadratic formula: $x = \frac{-(-2) \pm \sqrt{(-2)^2 - 4 \times 1 \times -3}}{2 \times 1} = \frac{2 \pm \sqrt{4+12}}{2} = \frac{2 \pm \sqrt{16}}{2}$. So, the solutions are $x = -1$ and $x = 3$.

5) For the given equation $4x^2 - 12x + 9 = 0$, $a = 4$, $b = -12$, and $c = 9$. Applying the quadratic formula: $x = \frac{-(-12) \pm \sqrt{(-12)^2 - 4 \times 4 \times 9}}{2 \times 4} = \frac{12 \pm \sqrt{144-144}}{8} = \frac{12}{8}$. So, the repeated solution is $x = \frac{3}{2}$.

6) The vertex of the parabola for the function $y = x^2 - 2x + 3$ is at $h = \frac{-(-2)}{2} = 1$ and $k = f(1) = 1^2 - 2 \times 1 + 3 = 2$. Therefore, the vertex is at $(1,2)$. The y-intercept is at $(0,3)$ since $f(0) = 0^2 - 2 \times 0 + 3 = 3$. This parabola opens upwards since the coefficient of x^2 is positive. Since the vertex is the minimum point of the parabola, there are no x-intercepts.

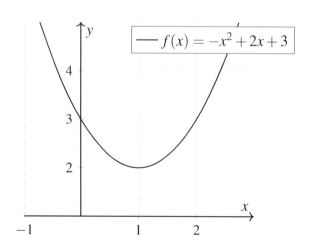

7) By examining the graph of this function, the vertex is the highest point as the parabola opens downwards, which is at $(1,1)$. The y-intercept is the point where the graph intersects the y-axis, which is at $(0,0)$. The x-intercepts are the points where the graph intersects the x-axis, which are at $(0,0)$ and $(2,0)$.

8) The vertex of the parabola for the function $y = 2x^2 + 4x + 1$ is at $h = \frac{-b}{2a} = \frac{-4}{2\times 2} = -1$ and $k = f(-1) = 2(-1)^2 + 4\times(-1) + 1 = -1$. Therefore, the vertex is at $(-1,-1)$. The y-intercept is at $(0,1)$ since $f(0) = 2(0)^2 + 4\times 0 + 1 = 1$. This parabola opens upwards, and since the parabola's vertex is its minimum point. The x-intercepts are $(\frac{-2+\sqrt{2}}{2}, 0)$ and $(-\frac{\sqrt{2}+1}{\sqrt{2}}, 0)$.

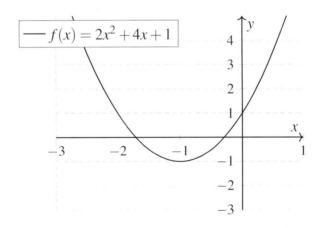

9) The orientation of the parabola is upwards since it opens in the upward direction. The vertex is at the lowest point, which is at $(-1,-4)$. The y-intercept is the point where the graph intersects the y-axis, which is at $(0,-3)$. The x-intercepts are where the graph intersects the x-axis, which are at $(-3,0)$ and $(1,0)$.

10) The vertex of the parabola for the function $y = -3x^2 + 6x + 5$ is at $h = \frac{-b}{2a} = \frac{-6}{2\times(-3)} = 1$ and $k = f(1) = -3(1)^2 + 6\times 1 + 5 = 8$. Therefore, the vertex is at $(1,8)$. The y-intercept is at $(0,5)$ since $f(0) = -3(0)^2 + 6\times 0 + 5 = 5$. This parabola opens downwards, the parabola's vertex is its maximum point and there are two x-intercepts.

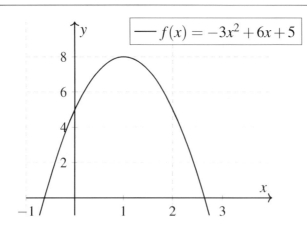

11) The numbers 2 and 5 add up to 7 and multiply to 10. So, the blanks should be filled with 7, to factorize the equation.

12) The equation factorizes to $(x+3)(x+6) = 0$, hence the roots are -3 and -6.

13) The equation $x^2 + 6x + 9 = 0$ can be factored into $(x+3)^2 = 0$ using the completing the square method.

14) To factorize the quadratic equation, the two numbers should add up to b and multiply to c.

15) The solutions of this quadratic equation are $x = -2, -2$.

16) Substitute $x = 0$, $x = 1$, and $x = -1$ into the given quadratic equation and solve for the y values: $f(0) = 0 + 0 + 4 = 4$, $f(1) = 1 + 4 + 4 = 9$, $f(-1) = 1 - 4 + 4 = 1$.

17) The domain of any quadratic function is $x \in \mathbb{R}$. For the range, the vertex of the function can be found using $-\frac{b}{2a} = -\frac{-4}{2} = 2$. Substituting $x = 2$, $f(2) = 2^2 - 4(2) + 4 = 0$, resulting in a range of $[0, +\infty)$.

18) The domain of any quadratic function is $x \in \mathbb{R}$. For the range, the vertex of the function can be found using $-\frac{b}{2a} = -\frac{6}{-4} = 1.5$. Substituting $x = 1.5$, $f(1.5) = -2(1.5)^2 + 6(1.5) - 3 = 1.5$, resulting in a range of $(-\infty, 1.5]$.

19) The domain of any quadratic function is $x \in \mathbb{R}$. For the range, the vertex of the function can be found using $-\frac{b}{2a} = -\frac{-12}{6} = 2$. Substituting $x = 2$, $f(2) = 3(2)^2 - 12(2) + 7 = -5$, resulting in a range of $[-5, +\infty)$.

20) The domain of any quadratic function is $x \in \mathbb{R}$. For the range, the vertex of the function can be found using $-\frac{b}{2a} = -\frac{4}{-2} = 2$. Substituting $x = 2$, $f(2) = -(2)^2 + 4(2) - 2 = 2$, resulting in a range of $(-\infty, 2]$.

7.10 Practices

21) The domain of any quadratic function is $x \in \mathbb{R}$. For the range, the vertex of the function can be found using $-\frac{b}{2a} = -\frac{-16}{8} = 2$. Substituting $x = 2$, $f(2) = 4(2)^2 - 16(2) + 9 = -7$, resulting in a range of $[-7, +\infty)$.

22) Arrange the algebra tiles to form a square with sides $x+3$ and $x+5$.

23) Arrange the algebra tiles to form a square with sides $x+7$ and $x+3$.

24) Arrange the algebra tiles to form a square with sides $x+2$ and $x+3$.

25) Arrange the algebra tiles to form a square with side $x+7$.

26) Arrange the algebra tiles to form a square with side $x+2$.

27) Using the equation $0 = a(x-h)^2 + k$ with $h=2$, $k=-3$, and $a=1$, we solve for a using the point $(4,1)$. This gives the final equation of the function as $f(x) = (x-2)^2 - 3$.

28) We use the vertex form of the quadratic function and substitute $h = -1$, $k = 5$, and the point $(2, -4)$ in to find a. After solving, we get $a = -1$. This gives the final equation as $f(x) = -(x+1)^2 + 5$.

29) Given the point when the ball hits the ground $(4, 0)$, we use this point and the vertex $(h, k) = (2, 10)$ and substitute into the equation $y = a(x-h)^2 + k$. Solving for a gives $a = -\frac{5}{2}$. This gives the final equation as $f(t) = -\frac{5}{2}(t-2)^2 + 10$.

30) We use the vertex form of quadratic function substituting $h = -1$, $k = 2$ and point $(1,6)$ in to find a. After simplifying we get $a = 1$. Thus, the equation is $f(x) = (x+1)^2 + 2$.

31) Given the point when the stone hits the ground $(3, 0)$, we substitute this point and the vertex $(h, k) = (1, 4)$ into the equation $y = a(x-h)^2 + k$. Solving for a gives $a = -1$. Hence, the resulting equation is $h(t) = -(t-1)^2 + 4$.

32) Since the parabola opens upwards and the vertex is at $(1, -4)$, the function is in the form $f(x) = a(x-h)^2 + k$, where $h = 1$ and $k = -4$. Also given that $f(0) = a(-1)^2 - 4 = -3$, we find that $a = 1$. So, the equation of the function is $f(x) = (x-1)^2 - 4$.

33) Given that the parabola opens downwards and the vertex is at $(2, 3)$, the function is in the form $f(x) = a(x-h)^2 + k$, where $h = 2$ and $k = 3$. Given that the function passes through the point $(1, 4)$, substituting these values into the equation yields $a = 1$. So, the equation is $f(x) = (x-2)^2 + 3$.

34) Since the vertex is at the origin and the parabola opens downwards, the function is in the form $f(x) = ax^2$. Substituting the point $(2, -4)$ into the equation, we find $a = -1$. Therefore, the equation is $f(x) = -x^2$.

35) Given that the parabola opens upwards and the vertex is at $(1, 2)$, the function is in the form $f(x) = a(x-h)^2 + k$, where $h = 1$ and $k = 2$. Substituting the y-intercept $f(0) = 1$ into the equation, we find $a = -1$. So, the equation of the function is $f(x) = -(x-1)^2 + 2$.

36) Since the vertex is at $(4, 2)$ and the parabola opens downwards, the function is in the form $f(x) = a(x-h)^2 + k$, where $h = 4$ and $k = 2$. Substituting the point $(-2, -10)$ into the equation gives us $a = -\frac{1}{3}$. So, the equation is $f(x) = -\frac{1}{3}(x-4)^2 + 2$.

8. Relations and Functions

8.1 Understanding Function Notation and Evaluation

A relation is simply a set of inputs and outputs, usually expressed as ordered pairs (x,y), where x is referred to as the input and y as the output. One of the unique characteristics of a relation is that it can have multiple outputs for an input.

Key Point

Functions are mathematical operations that assign unique outputs to given inputs. Here, instead of using y, we use $f(x)$, where x is the input and $f(x)$ is the output of the function.

Function notation is a way of representing a function that is derived from an equation. Mostly, functions make use of the letter f, although any letter can be utilized. Writing a function as $f(x)$ instead of y emphasizes that the function's value depends solely on x.

Key Point

When we evaluate functions, we substitute specific values of x into the function formula to find the corresponding output, $f(x)$.

Example Suppose we have a function $f(x) = 2x+3$ and we want to evaluate $f(1)$.

Solution: To solve for $f(1)$, we simply replace x in the function's rule with the value 1: $f(1) = 2(1)+3 = 5$.

 **Example** If we have $f(x) = 2x^2 + 3x - 1$, find $f(-1)$.

Solution: Substitute $x = -1$ into the equation: $f(-1) = 2(-1)^2 + 3(-1) - 1 = 2 - 3 - 1 = -2$.

8.2 Completing a Function Table from an Equation

A function table is simply a table that showcases this relationship for distinct inputs and their corresponding outputs under the function operation.

> A function table helps us understand the behavior of a function at different input (x) values by listing down the corresponding output (y) values.

Simply put, all you need to do is substitute each x-value in the function's equation and solve for y, the output.

 **Example** Given the function $f(x) = 2x - 3$, complete the function table:

x	1	2	3
$f(x)$			

Solution: Replacing values of x into the function equation, we get:

For $x = 1$, $f(x) = 2(1) - 3 = -1$.

For $x = 2$, $f(x) = 2(2) - 3 = 1$.

For $x = 3$, $f(x) = 2(3) - 3 = 3$.

So, the completed function table is:

x	1	2	3
$f(x)$	-1	1	3

The graph of this function is shown in the following figure:

8.2 Completing a Function Table from an Equation

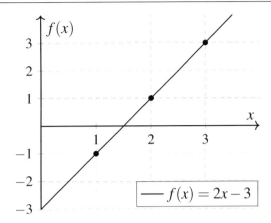

 Example   Complete the function table for $f(x) = x^2 + 2$

x	-1	0	1
$f(x)$			

Solution: Substituting values of x into the function equation, we get:

For $x = -1$, $f(-1) = (-1)^2 + 2 = 3$.

For $x = 0$, $f(0) = (0)^2 + 2 = 2$.

For $x = 1$, $f(1) = (1)^2 + 2 = 3$.

Therefore, the function table becomes:

x	-1	0	1
$f(x)$	3	2	3

The graph of this function is shown in the following figure:

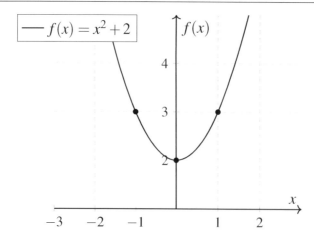

8.3 Determining Domain and Range of Relations

The domain of a relation is the set of all possible inputs (usually *x*-values), while the range is the collection of all possible outputs (usually *y*-values).

> **Key Point**
>
> The domain of a relation are the *x*-values (or inputs), while the range are the *y*-values (or outputs).

Example Determine the domain and range of the relation $\{(1,2),(3,4),(5,6)\}$.

Solution: The domain of this relation is the set of all *x*-values. So, the domain is $\{1,3,5\}$.

The range of this relation is the set of all *y*-values. Therefore the range is $\{2,4,6\}$.

Example Determine the domain and range of the relation $\{(7,8),(7,10),(2,10)\}$.

Solution: The domain of this relation is the set of all *x*-values. So, the domain is $\{7,2\}$.

The range of this relation is the set of all *y*-values. Therefore the range is $\{8,10\}$.

Note: Despite 7 and 10 being repeated in the pairs, they each only appear once in the domain and range.

8.4 Performing Addition and Subtraction of Functions

Consider two functions $f(x)$ and $g(x)$. When adding these functions together, we add the results of each function at each input. This means that to find $(f+g)(x)$, we add together $(f(x)+g(x))$.

Similarly, if we need to subtract a function $g(x)$ from another function $f(x)$, we subtract $g(x)$ from $f(x)$ at

8.5 Performing Multiplication and Division of Functions

every value of x. This operation creates the new function $(f-g)(x)$, where $(f-g)(x) = f(x) - g(x)$.

Key Point

> The addition $(f+g)(x)$ or subtraction $(f-g)(x)$ of two functions gives us a new function where addition or subtraction has been performed at each value of x.

Note that $(f-g)(x)$ is not equal to $(g-f)(x)$.

Example Suppose we have $f(x) = x^2$ and $g(x) = 2x+3$. Find $(f+g)(x)$ and $(f-g)(x)$.

Solution: For $(f+g)(x)$, we simply add $f(x)$ and $g(x)$ together: $(f+g)(x) = f(x) + g(x) = x^2 + 2x + 3$.

For $(f-g)(x)$, we subtract $g(x)$ from $f(x)$: $(f-g)(x) = f(x) - g(x) = x^2 - (2x+3) = x^2 - 2x - 3$.

Example For the functions $f(x) = 3x - 1$ and $g(x) = -x+2$, calculate $(f-g)(1)$ and $(g-f)(1)$.

Solution: To find $(f-g)(1)$, we first find $f(1)$ and $g(1)$, and then subtract them: $(f-g)(1) = f(1) - g(1) = (3 \times 1 - 1) - (-1 \times 1 + 2) = 2 - 1 = 1$.

For $(g-f)(1)$, following the same steps, we have: $(g-f)(1) = g(1) - f(1) = (-1 \times 1 + 2) - (3 \times 1 - 1) = 1 - 2 = -1$.

8.5 Performing Multiplication and Division of Functions

For the multiplication of functions, if we need to multiply $f(x)$ and $g(x)$, the result will be a new function, say $(f \times g)(x)$, where $(f \times g)(x) = f(x) \times g(x)$.

In a similar vein, if we are to divide a function $f(x)$ by $g(x)$, the new function denoted as $(\frac{f}{g})(x)$ will be defined as $(\frac{f}{g})(x) = \frac{f(x)}{g(x)}$. However, note that we can only perform this operation if $g(x) \neq 0$.

Key Point

> Multiplication $(f \times g)(x)$ or division $(\frac{f}{g})(x)$ (for $g(x) \neq 0$) of two functions gives us a new function where multiplication or division has been performed at each functional value.

Note that, like subtraction, division of functions is not commutative.

Example Suppose we have $f(x) = x^2$ and $g(x) = 2x+3$. Compute $(f \times g)(x)$ and $(\frac{f}{g})(x)$.

Solution: For $(f \times g)(x)$, we simply multiply $f(x)$ and $g(x)$ together: $(f \times g)(x) = f(x) \times g(x) = x^2 \times (2x+3) = 2x^3 + 3x^2$.

For $(\frac{f}{g})(x)$, we divide $f(x)$ by $g(x)$, assuming that $g(x) \neq 0$: $(\frac{f}{g})(x) = \frac{f(x)}{g(x)} = \frac{x^2}{2x+3}$.

Example For the functions $f(x) = 3x^2 - 1$ and $g(x) = x+2$, compute $(f \times g)(2)$ and $(\frac{f}{g})(1)$.

Solution: To find $(f \times g)(2)$, we first find $f(2)$ and $g(2)$, and then multiply them: $(f \times g)(2) = f(2) \times g(2) = (3 \times 2^2 - 1) \times (2+2) = 44$.

For $(\frac{f}{g})(1)$, we divide $f(1)$ by $g(1)$: $(\frac{f}{g})(1) = \frac{f(1)}{g(1)} = \frac{3 \times 1 - 1}{1+2} = \frac{2}{3}$.

8.6 Composing Functions

The function composition is a type of operation where the output of one function becomes the input of another. This operation creates a new function, named a composite function.

The composition of two functions, $f(x)$ and $g(x)$, is denoted as $(f \circ g)(x)$, you first evaluate $g(x)$ and then the resulting output becomes the input into the f function.

Function composition $(f \circ g)(x)$ is a process where the output of one function, $g(x)$, is used as the input for another function, $f(x)$.

Example Given the functions $f(x) = x^2 + 1$ and $g(x) = 2x + 3$, find $(f \circ g)(x)$ and $(g \circ f)(x)$.

Solution: To find $(f \circ g)(x)$, we substitute $g(x)$ into $f(x)$. Thus: $(f \circ g)(x) = f(g(x)) = (2x+3)^2 + 1 = 4x^2 + 12x + 10$.

To find $(g \circ f)(x)$, we substitute $f(x)$ into $g(x)$. Hence: $(g \circ f)(x) = g(f(x)) = 2(x^2 + 1) + 3 = 2x^2 + 5$.

Example Given $f(x) = 2x - 1$ and $g(x) = x^2 + 4$, compute $(f \circ g)(1)$ and $(g \circ f)(2)$.

Solution: To compute $(f \circ g)(1)$, first find $g(1)$, then substitute this into $f(x)$. Hence: $(f \circ g)(1) = f(g(1)) = f(1^2 + 4) = f(5) = 2 \times 5 - 1 = 9$.

To compute $(g \circ f)(2)$, first find $f(2)$, then substitute this value into $g(x)$: $(g \circ f)(2) = g(f(2)) = g(2 \times 2 - 1) = g(3) = 3^2 + 4 = 13$.

8.7 Evaluating Exponential Functions

Exponential functions are a type of mathematical function in which the argument, or input, is the exponent. These functions represent various forms of growth and decay, including population growth, radioactive decay, and compound interest.

An exponential function can be recognized in the form $f(x) = a \cdot b^x$, where a is a non-zero real number and b is a positive real number such that $b \neq 1$.

To evaluate an exponential function means to find its value for a specific value of x. This can generally be done by simply substituting the specified x-value into the function and calculating the result accordingly.

> **Key Point**
>
> Calculating the value of an exponential function involves substituting the input into the exponent of the base.

 **Example** Given the function $f(x) = 2^x$, find the value of $f(3)$.

Solution: Substitute the given x-value into the function: $f(3) = 2^3 = 8$.

 **Example** Given the function $g(x) = 5(3^x)$, find the value of $g(2)$.

Solution: To find the value of $g(2)$, we just need to substitute 2 into the function: $g(2) = 5 \times (3^2) = 5 \times 9 = 45$.

8.8 Matching Exponential Functions with Graphs

An exponential function is generally of the form $f(x) = a \cdot b^x$, where a and b are constants, and x is the variable. The graph of an exponential function exhibits a unique 'J' shape and can further be broken down into two categories:

1. For $0 < b < 1$, the graph represents an exponential decay, starting from a high point and gradually reducing as it approaches the x-axis.

2. For $b > 1$, the graph shows an exponential growth situation that starts from a relatively low point and increases rapidly.

A crucial component of an exponential function's graph is the y-intercept. The y-intercept is where the graph intersects the y-axis. For our exponential function $f(x) = a \cdot b^x$, the y-intercept is a.

Key Point

An exponential function $f(x) = a \cdot b^x$ displays a 'J' shaped graph. It exhibits exponential growth for $b > 1$ and decay for $0 < b < 1$. The y-intercept of the graph is at a.

Example
Find the exponential function which matches the graph below. The y-intercept is 4 and the base is greater than 1.

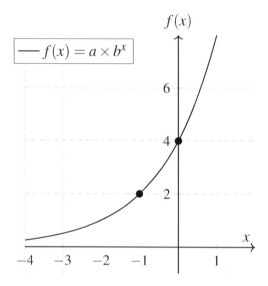

Via the y-intercept, we know that $a = 4$. Since the base is greater than 1, it tells us that the graph shows an exponential growth. Thus, we can write the function as $f(x) = 4 \times b^x$. The point $(-1, 2)$ is on the graph, thus satisfies in the equation of graph. So, we have $2 = 4 \times b^{-1}$. Solving for b we get: $b = 2$. Therefore, the exponential function is $f(x) = 4 \cdot 2^x$.

Example
Find the exponential function which matches the graph below. The y-intercept is 2 and the base is less than 1.

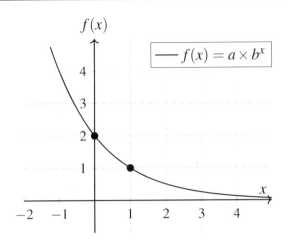

Here, we discern that $a = 2$ from the y-intercept. Since the base is less than 1, it tells us that the graph represents an exponential decay. Hence, we can create the function $f(x) = 2 \times b^x$. By considering the point $(1, 1)$, we have: $f(1) = 1$ and hence, $1 = 2 \times b^1$. Solving for b we get: $b = \frac{1}{2}$. Therefore, the exponential function which matches the graph is $f(x) = 2 \cdot \left(\frac{1}{2}\right)^x$.

8.9 Writing Exponential Functions from Word Problems

Writing an exponential function from a word problem involves extracting the relevant details, identifying the constant parameters, and forming the equation.

We know that a generic exponential function is of the form $f(x) = a \cdot b^x$, where:

- 'a' is the initial or starting value,
- 'b' is the rate of growth or decay (depending on whether it is greater than or less than 1), and
- 'x' is the variable.

Key Point

The initial value in a word problem usually signifies the y-intercept 'a'. The rate of growth or decay identified in the problem is the base 'b'.

Example

Suppose a new phone app is released, and the number of downloads doubles every day. On the first day of release, the app was downloaded 50 times. Write an exponential function to represent the number of downloads, 'd', after 'x' days.

Solution: In this problem, we identify:

1. The initial value, 'a', is 50, as the app was downloaded 50 times initially.
2. The rate of change is doubling, so 'b' = 2 (i.e., growth).

Thus, the function becomes $d(x) = 50 \times 2^x$.

Example A certain radioactive material decreases by 5% each year. If the initial quantity of the material was 100 kg, determine the exponential function that models this decay.

Solution: From the problem, we identify:

1. The initial value, 'a', is 100, as there were initially 100 kg of the material.
2. The material decreases by 5% every year. This is a decay percentage, so 'b' = $1 - 0.05 = 0.95$.

Hence, the decay function becomes $m(x) = 100 \times 0.95^x$, with $m(x)$ representing the material's quantity after x years.

8.10 Understanding Function Inverses

In the real world, we often come across situations where we find the concept of inverse very handy. Think of it as a reverse operation. If multiplication is an operation, division is its inverse.

A similar concept applies to functions in algebra. If we have a function that takes us from A to B, the inverse of that function takes us back from B to A.

Mathematically, a function $f(x)$ has an inverse $f^{-1}(x)$ if for every x in the domain of $f(x)$, $f^{-1}(f(x)) = x$ and for every x in the domain of $f^{-1}(x)$, $f(f^{-1}(x)) = x$. Essentially, replacing $f(x)$ with y, the inverse function swaps the roles of x and y.

Key Point

> To find the inverse of $f(x)$, replace $f(x)$ with y, then interchange x and y and solve the resulting equation for y. Remember not every function has an inverse. At least, it should be a one-to-one function which means that every element in the domain is mapped to a unique element in the range.

We can also understand function inverses from a graphical viewpoint. The graph of the inverse function is a reflection of the graph of the original function over the line $y = x$.

8.10 Understanding Function Inverses

Example Find the inverse of the function $f(x) = 2x + 3$.

Solution: First, replace $f(x)$ with y, giving us the equation $y = 2x + 3$.

Then, interchange x and y to get $x = 2y + 3$.

Solving this equation for y, we get $y = \frac{x-3}{2}$.

Therefore, the inverse of the function $f(x) = 2x + 3$ is $f^{-1}(x) = \frac{x-3}{2}$.

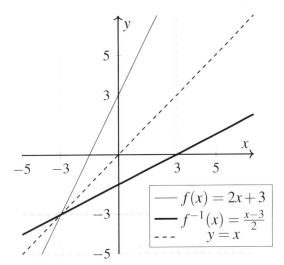

Example From a graphical viewpoint, show that the function $f(x) = x^2$ and its inverse $f^{-1}(x) = \sqrt{x}$ are reflections of each other in the line $y = x$ and interval $[0, +\infty)$.

Solution: To do this, we can plot both functions on the same set of axes.

Plotting $f(x) = x^2$, we get a parabola opening upwards.

Plotting $f^{-1}(x) = \sqrt{x}$, we get the half of the parabola $y = x^2$ above the line $y = x$, which looks like a reflection of the bottom half of the parabola in the line $y = x$.

Therefore, we can visually confirm that $f(x) = x^2$ and $f^{-1}(x) = \sqrt{x}$ are reflections of each other in the line $y = x$.

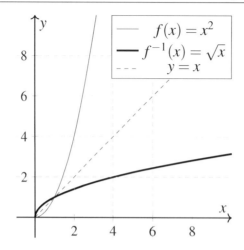

8.11 Understanding Rate of Change and Slope

Rate of change and *slope* are key concepts associated with functions, specifically linear functions, guiding us in describing their behavior.

The 'rate of change' in a function refers to the speed at which the y values (output values) change for every single unit change in the x values (input values). It provides an essential glimpse into the behavior of the function over its entire domain.

Meanwhile, in a linear function, the 'slope' is the constant rate of change, and it indicates the steepness of the line.

Key Point

For every function we have:

$$\text{Rate of Change} = \frac{\text{Change in Output}(Y)}{\text{Change in Input}(X)} = \frac{\Delta Y}{\Delta X}.$$

For linear functions, rate of change is the slope of the line.

Example

Given the function $f(x) = 3x + 2$. What is the rate of change of this function?

Solution: For a linear function in the form $f(x) = mx + b$, where m is the slope and b is the y-intercept, the number associated with x is the slope or the rate of change. In this function, 3 is associated with x. Therefore, the rate of change or slope of $f(x) = 3x + 2$ is 3.

8.12 Practices

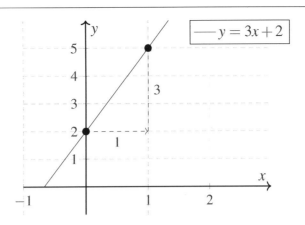

Example Find the rate of change between the points $(4,-2)$ and $(2,6)$ on the line.

Solution: The rate of change of a linear function is the slope of the line. We can calculate the slope using the formula: $\frac{\Delta y}{\Delta x} = \frac{y_2 - y_1}{x_2 - x_1}$.

Substituting these points into the formula, we get: Slope $= \frac{6-(-2)}{2-4} = \frac{8}{-2} = -4$.

So, the rate of change between the points $(4,-2)$ and $(2,6)$ is -4.

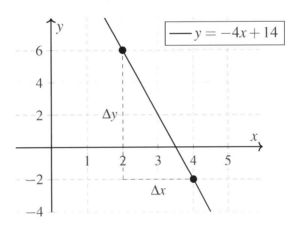

8.12 Practices

Solve:

1) Given that $f(x) = x^2 + 3x + 2$ and $g(x) = -x^2 + 2x - 1$, find $(f+g)(-1)$ and $(f-g)(0)$.

2) If $f(x) = 2x - 3$ and $g(x) = x^2 + 2$, find $(f+g)(2)$ and $(f-g)(-1)$.

3) Given $f(x) = 3x^2 - 2x + 1$ and $g(x) = -x^2 + x - 2$, compute $(f+g)(-2)$ and $(f-g)(3)$.

4) If $f(x) = 2x^3 + x^2 - x + 1$ and $g(x) = -x^3 + x^2 + x - 2$, find $(f+g)(-1)$ and $(f-g)(1)$.

5) Given $f(x) = 4x - 2$ and $g(x) = -2x + 3$, compute $(f+g)(1)$ and $(f-g)(-1)$.

Solve:

6) Given $f(x) = 3x^2 + 2$ and $g(x) = x - 1$, compute $(f \times g)$ and $\left(\frac{f}{g}\right)$ for $x = 2$.

7) Given $f(x) = 2x + 3$ and $g(x) = x^2 - 1$, compute $(f \times g)$ and $\left(\frac{f}{g}\right)$ for $x = -1$.

8) Given $f(x) = x + 4$ and $g(x) = x^3$, compute $(f \times g)$ and $\left(\frac{f}{g}\right)$ for $x = 1$.

9) Given $f(x) = x^2 - 3x + 2$ and $g(x) = x^2 + 2x - 1$, compute $(f \times g)$ and $\left(\frac{f}{g}\right)$ for $x = 0$.

10) Given $f(x) = 2x^3 - x^2 + 1$ and $g(x) = x + 2$, compute $(f \times g)$ and $\left(\frac{f}{g}\right)$ for $x = -1$.

True or False:

11) For any functions f and g, $(f \circ g)(x) = (g \circ f)(x)$. True or False?

12) For any functions f and g, $(g \circ f)(x)$ means $f(x)$ is substituted into $g(x)$. True or False?

13) If $f(x) = 2x + 1$ and $g(x) = x^2$, then $(f \circ g)(x) = 2x^2 + 1$. True or False?

14) If $f(x) = 3x - 1$ and $g(x) = x + 4$, then $(g \circ f)(2) = 10$. True or False?

15) Given $f(x) = x^2$ and $g(x) = x + 5$, then $f(g(3)) = 64$. True or False?

Evaluate exponential functions:

16) Given the function $f(x) = 4^x$, find the value of $f(2)$.

17) Given the function $g(x) = 3 \cdot (2^x)$, find the value of $g(4)$.

18) Given the function $h(x) = 5^x$, find the value of $h(3)$.

19) Given the function $j(x) = 7 \cdot (3^x)$, find the value of $j(2)$.

8.12 Practices

20) Given the function $k(x) = 6^x$, find the value of $k(1)$.

Fill in the Blank:

21) A population of bacteria triples every hour. If the initial population was 500, the exponential function representing the bacteria population, 'p', after 'x' hours will be $p(x) = 500 \cdot $ _____^{x}.

22) A car's value depreciates by 15% every year. If the car originally cost $20000, the exponential function modelling this depreciation, 'v', after 'x' years is $v(x) = 20000 \cdot $ _____^{x}.

23) A certain radioactive substance decreases by 7% each month. If the initial quantity of this substance was 80 grams, the exponential function representing this decay, 'r', after 'x' months is $r(x) = $ _____ · _____^{x}.

24) A new software gets downloaded five times more every day. If there were 20 downloads on the first day, the exponential function for the 'd', number of downloads, after 'x' days is $d(x) = $ _____ · _____^{x}.

25) A type of fungus doubles its mass every 3 days. If the initial mass of the fungus was 2g, the function representing the fungal mass, 'm', after 'x' days would be $m(x) = $ _____ · _____^{x}.

Solve:

26) Find the inverse of the function $f(x) = 7x - 5$.

27) Find the inverse of the function $f(x) = \frac{3}{x}$.

28) Find the inverse of the function $f(x) = \sqrt{x+2}$.

29) Find the inverse of the function $f(x) = x^3$.

30) Find the inverse of the function $f(x) = \frac{1}{x} + 4$.

True or False:

31) The domain and range of the relation $\{(4,7),(4,8),(4,9)\}$ are $\{4\}$ and $\{7,8,9\}$ respectively. True or false?

32) The relation $\{(1,2),(1,3),(1,4)\}$ has a domain $\{1,2,3,4\}$ and a range $\{1\}$. True or false?

33) For the relation $\{(4,5),(4,6)\}$, the domain is $\{4\}$ and the range is $\{5,6\}$. True or false?

34) The domain is made up of the *y*-values in a relation, and the range is formed by the *x*-values. True or false?

35) The domain and range for the relation $\{(7,8),(9,8),(10,8)\}$ are $\{7,9,10\}$ and $\{8\}$ respectively. True or false?

Solve:

36) If given $g(x) = 7x - 4$, find the slope or rate of change.

37) Find the rate of change between the points $(5,3)$ and $(2,-1)$.

38) If a line passes through the points $(-6,-4)$ and $(2,4)$, find the rate of change.

39) Calculate the slope for the linear function $k(x) = -\frac{3}{2}x + 1$.

40) Find the rate of change between the points $(3,8)$ and $(-2,-4)$.

Solve:

41) Solve for f(x) when $f(x) = x^3 - 2x^2 - 5x - 6$ and $x = 3$.

42) Find $f(x)$ for $f(x) = 2x\sqrt{x} - x$ and $x = 4$.

43) Solve for $f(x)$ when $f(x) = x^2 + 4x - 12$ and $x = -2$.

44) Calculate $f(x)$ for $f(x) = \frac{x}{2} + 3$ and $x = 4$.

45) Solve for $f(x)$ when $f(x) = x^3 - 3x^2 + 2x - 1$ and $x = -1$.

8.12 Practices

Answer Keys

1) $(f+g)(-1) = -4$, $(f-g)(0) = 3$
2) $(f+g)(2) = 7$, $(f-g)(-1) = -8$
3) $(f+g)(-2) = 9$, $(f-g)(3) = 30$
4) $(f+g)(-1) = 0$, $(f-g)(1) = 4$
5) $(f+g)(1) = 3$, $(f-g)(-1) = -11$
6) $(f \times g)(2) = 14$, $(\frac{f}{g})(2) = 14$
7) $(f \times g)(-1) = 0$, $(\frac{f}{g})(-1)$ undefined
8) $(f \times g)(1) = 5$, $(\frac{f}{g})(1) = 5$
9) $(f \times g)(0) = -2$, $(\frac{f}{g})(0) = -2$
10) $(f \times g)(-1) = -2$, $(\frac{f}{g})(-1) = -2$
11) False
12) True
13) True
14) False
15) True
16) 16
17) 48
18) 125
19) 63
20) 6
21) 3
22) 0.85
23) 80, 0.93
24) 20, 5
25) 2, $\sqrt[3]{2}$
26) $f^{-1}(x) = \frac{x+5}{7}$
27) $f^{-1}(x) = \frac{3}{x}$
28) $f^{-1}(x) = x^2 - 2$
29) $f^{-1}(x) = \sqrt[3]{x}$
30) $f^{-1}(x) = \frac{1}{x-4}$
31) True
32) False
33) True
34) False
35) True
36) 7
37) $\frac{4}{3}$
38) 1
39) -1.5
40) 2.4
41) -12
42) 12
43) -16
44) 5
45) -7

Answers with Explanation

1) $(f+g)(-1) = f(-1) + g(-1) = (-1)^2 + 3 \times (-1) + 2 + -(-1)^2 + 2 \times (-1) - 1 = -4$.
$(f-g)(0) = f(0) - g(0) = 0^2 + 3 \times 0 + 2 - (-(0^2) + 2 \times 0 - 1) = 3$.

2) $(f+g)(2) = f(2) + g(2) = 2 \times 2 - 3 + 2^2 + 2 = 7$.
$(f-g)(-1) = f(-1) - g(-1) = 2 \times (-1) - 3 - ((-1)^2 + 2) = -8$.

3) $(f+g)(-2) = f(-2) + g(-2) = 3 \times (-2)^2 - 2 \times (-2) + 1 + -(-2)^2 + -2 - 2 = 9$.
$(f-g)(3) = f(3) - g(3) = 3 \times 3^2 - 2 \times 3 + 1 - (-(3)^2 + 3 - 2) = 22 + 8 = 30$.

4) $(f+g)(-1) = f(-1) + g(-1) = 2 \times (-1)^3 + (-1)^2 - (-1) + 1 + -(-1)^3 + (-1)^2 + -1 - 2 = 0$.
$(f-g)(1) = f(1) - g(1) = 2 \times 1^3 + 1^2 - 1 + 1 - (-(1)^3 + 1^2 + 1 - 2) = 3 + 1 = 4$.

5) $(f+g)(1) = f(1) + g(1) = 4 \times 1 - 2 + -2 \times 1 + 3 = 4 - 2 - 2 + 3 = 3$.
$(f-g)(-1) = f(-1) - g(-1) = 4 \times (-1) - 2 - (-2 \times (-1) + 3) = -4 - 2 - (2 + 3) = -6 - 5 = -11$.

6) For $(f \times g)(2)$, we first find $f(2) = 3 \times 2^2 + 2 = 14$ and $g(2) = 2 - 1 = 1$. So, $(f \times g)(2) = f(2) \times g(2) = 14 \times 1 = 14$. For $(\frac{f}{g})(2)$, we get $(\frac{f}{g})(2) = \frac{f(2)}{g(2)} = \frac{14}{1} = 14$, considering $g(2) \neq 0$.

7) For $(f \times g)(-1)$, we first find $f(-1) = 2 \times (-1) + 3 = 1$ and $g(-1) = (-1)^2 - 1 = 0$. So, $(f \times g)(-1) = f(-1) \times g(-1) = 1 \times 0 = 0$. For $(\frac{f}{g})(-1)$, we get $(\frac{f}{g})(-1) = \frac{f(-1)}{g(-1)} = \frac{1}{0}$. As division by zero is undefined, $(\frac{f}{g})(-1)$ is undefined.

8) For $(f \times g)(1)$, we first find $f(1) = 1 + 4 = 5$ and $g(1) = 1^3 = 1$. So, $(f \times g)(1) = f(1) \times g(1) = 5 \times 1 = 5$. For $(\frac{f}{g})(1)$, we get $(\frac{f}{g})(1) = \frac{f(1)}{g(1)} = \frac{5}{1} = 5$.

9) For $(f \times g)(0)$, we first find $f(0) = 0^2 - 3 \times 0 + 2 = 2$ and $g(0) = 0^2 + 2 \times 0 - 1 = -1$. So, $(f \times g)(0) = f(0) \times g(0) = 2 \times -1 = -2$. For $(\frac{f}{g})(0)$, we get $(\frac{f}{g})(0) = \frac{f(0)}{g(0)} = \frac{2}{-1} = -2$.

10) For $(f \times g)(-1)$, we first find $f(-1) = 2 \times (-1)^3 - (-1)^2 + 1 = -2$ and $g(-1) = -1 + 2 = 1$. So, $(f \times g)(-1) = f(-1) \times g(-1) = -2 \times 1 = -2$. For $(\frac{f}{g})(-1)$, we get $(\frac{f}{g})(-1) = \frac{f(-1)}{g(-1)} = -2$.

8.12 Practices

11) Ordinarily, the order of function composition matters: $(f \circ g)(x)$ and $(g \circ f)(x)$ generally give different results.

12) By definition, $(g \circ f)(x) = g(f(x))$, involves substituting $f(x)$ into the function g.

13) Substitute $g(x)$ into $f(x)$ to produce: $(f \circ g)(x) = f(g(x)) = f(x^2) = 2x^2 + 1$.

14) First, compute $f(2) = 3 \times 2 - 1 = 5$. Now substitute the value of $f(2)$ in $g(x)$ to get $g(f(2)) = g(5) = 5 + 4 = 9$, which is not equal to 10. Thus the statement is False.

15) Calculate $g(3)$ first, which gives $g(3) = 3 + 5 = 8$. Now, substitute 8 into $f(x)$ to get $f(g(3)) = f(8) = 8^2 = 64$. Hence, the statement is true.

16) To evaluate $f(2)$, we substitute 2 for x in the function: $f(2) = 4^2 = 16$.

17) To evaluate $g(4)$, we substitute 4 for x in the function: $g(4) = 3 \cdot (2^4) = 3 \times 16 = 48$.

18) To find $h(3)$, we simply substitute 3 for x in the function: $h(3) = 5^3 = 125$.

19) To evaluate $j(2)$, we substitute 2 for x in the function: $j(2) = 7 \cdot (3^2) = 7 \times 9 = 63$.

20) To evaluate $k(1)$, we substitute 1 for x in the function: $k(1) = 6^1 = 6$.

21) The bacteria population triples every hour so we can fill in the blank with 3.

22) The car value depreciates by 15% every year. So $b = 1 - 0.15 = 0.85$.

23) The initial quantity is 80, so we can fill in the first blank with 80. Moreover, the substance decreases by 7% each month which means $b = 1 - 0.07 = 0.93$. Thus, we fill in the second blank with 0.93.

24) The rate of growth is multiplying by 5 every day, which gives us $b = 5$ for the second blank. The initial number of downloads is 20, so we fill in the first blank with 20.

25) The initial mass of the fungus is 2g, so we fill in the first blank with 2. The fungus doubles its mass every 3 days, which implies that $b = \sqrt[3]{2}$ and we fill in the second blank with $\sqrt[3]{2}$.

26) Replace $f(x)$ with y to get $y = 7x - 5$. Swap x and y to get $x = 7y - 5$. Solving for y, we find $f^{-1}(x) = \frac{x+5}{7}$.

27) Here $f(x) = \frac{3}{x}$ already satisfies the condition for the inverse of a function, check $f(f^{-1}(x)) = x$ holds. Hence, the inverse is $f^{-1}(x) = \frac{3}{x}$.

28) Replace $f(x)$ with y to get $y = \sqrt{x+2}$. Swap x and y to get $x = \sqrt{y+2}$. Solving for y, we find $f^{-1}(x) = x^2 - 2$.

29) To find the inverse, we first write the function as $y = x^3$. Next, we swap x and y to get $x = y^3$. Solving for y, we find $y = \sqrt[3]{x}$. Therefore, the inverse function is $f^{-1}(x) = \sqrt[3]{x}$.

30) Replace $f(x)$ with y to get $y = \frac{1}{x} + 4$. Swap x and y to get $x = \frac{1}{y} + 4$. Solving for y, we find $f^{-1}(x) = \frac{1}{x-4}$.

31) The statement is true because the single x-value 4 forms the domain and the y-values form the range $\{7, 8, 9\}$.

32) This is false since the domain is formed by the x-values, which is $\{1\}$, and the range is formed by the y-values, which are $\{2, 3, 4\}$.

33) This is true as the domain of a relation is made up of the x-values, which is $\{4\}$, and the range is formed by the y-values, which are $\{5, 6\}$.

34) This is false because the domain of a relation is made up of the x-values (inputs) and the range is formed by the y-values (outputs).

35) The statement is true: the domain is formed by the x-values, which is $\{7, 9, 10\}$, and the range is formed by the y-value, which is $\{8\}$.

36) For linear functions in the form $f(x) = mx + b$, the coefficient of x represents the slope or rate of change. Here, the slope or rate of change is 7.

37) The rate of change or slope can be found using the formula: Slope $= \frac{y_2 - y_1}{x_2 - x_1}$. By substituting the given points into the formula, we get: Slope $= \frac{3-(-1)}{5-2} = \frac{4}{3}$.

38) Again, by using the slope formula: Slope $= \frac{y_2 - y_1}{x_2 - x_1}$. And substituting for the points given: Slope $= \frac{4-(-4)}{2-(-6)} = \frac{8}{8} = 1$.

39) For linear functions in the form $f(x) = mx + b$, m represents the slope or rate of change. In this case, the

8.12 Practices

slope is $-\frac{3}{2} = -1.5$.

40) By using the slope formula: Slope = $\frac{y_2-y_1}{x_2-x_1}$. And substituting for the points given: Slope = $\frac{8-(-4)}{3-(-2)}$ = $\frac{12}{5} = 2.4$.

41) $f(x) = (3)^3 - 2(3)^2 - 5(3) - 6 = 27 - 18 - 15 - 6 = -12$.

42) $f(x) = 2 \times 4\sqrt{4} - 4 = 2 \times 4 \times 2 - 4 = 16 - 4 = 12$.

43) $f(x) = (-2)^2 + 4(-2) - 12 = 4 - 8 - 12 = -16$.

44) $f(x) = \frac{4}{2} + 3 = 2 + 3 = 5$.

45) $f(x) = (-1)^3 - 3(-1)^2 + 2(-1) - 1 = -1 - 3 - 2 - 1 = -7$.

9. Radical Expressions

9.1 Simplifying Radical Expressions

A radical expression contains a square root, cube root, or any other root, denoted by $\sqrt[n]{k}$. This symbol signifies finding the number that, when raised to the power n, yields the number inside the root (k). The term under the radical is the radicand, and n is the index.

Simplifying radical expressions involves expressing them in less complex equivalent forms without changing their values.

Key Point

To simplify a radical expression, find the prime factors of the radicand and use radical properties:
$$\sqrt[n]{x^a} = x^{\frac{a}{n}}, \quad \sqrt[n]{xy} = \sqrt[n]{x} \times \sqrt[n]{y}, \quad \sqrt[n]{\frac{x}{y}} = \frac{\sqrt[n]{x}}{\sqrt[n]{y}}.$$

Example Simplify $\sqrt{144x^2}$.

Solution: Factor $144x^2$ into $12^2 x^2$. Applying the radical rule $\sqrt{x^2} = x$, we get $\sqrt{144x^2} = \sqrt{12^2} \times \sqrt{x^2} = 12x$.

Example Simplify $\sqrt{50}$.

Solution: Before simplifying, we have to factor 50. We get 2 and 25, and 25 can be factored further to 5 and 5. Hence, the expression becomes: $\sqrt{50} = \sqrt{2 \times 5 \times 5} = \sqrt{2 \times (5)^2}$.

By the law of radicals, we can express this as: $\sqrt{5^2} \times \sqrt{2}$. So, $\sqrt{50}$ simplifies to $5\sqrt{2}$.

9.2 Performing Addition and Subtraction of Radical Expressions

The main concept in addition and subtraction of radical expressions revolves around the idea of like and unlike radicals.

Essentially, it is similar to how we can only add or subtract similar terms like $3x + 2x$ or $5y - 2y$, the same rule applies to radical expressions. If the radicand and index are the same, then we can add or subtract the radicals.

> **Key Point**
>
> Like radicals are radical expressions that have the same radicand and index. Addition and subtraction operations are only valid between like radicals.

Example Simplify $\sqrt{2} + \sqrt{2}$.

Solution: These are like radicals because both their radicands (the number under the radical) and their indices (the root number) are the same. The operation becomes straightforward once you think of the radicals as variables. This sum then becomes: $2\sqrt{2}$.

Example Simplify $3\sqrt{2} - 2\sqrt{3}$.

Solution: In this case, we have unlike radicals as they feature different radicands and the same index. Similar to variables, you cannot subtract $2y$ from $3x$. Hence, this expression cannot be simplified further and we leave it as is: $3\sqrt{2} - 2\sqrt{3}$.

9.3 Performing Multiplication of Radical Expressions

Multiplying radical expressions adheres to the same principles as with any algebraic expressions, with the essential understanding of the product rule for radicals.

Key Point

The product rule for radicals states:

Given $a, b \geq 0$ and n is a positive even integer, then $\sqrt[n]{a}\sqrt[n]{b} = \sqrt[n]{ab}$.

Given a, b are real numbers and n is a positive odd integer, then $\sqrt[n]{a}\sqrt[n]{b} = \sqrt[n]{ab}$.

Applying the product rule is a straightforward process, especially if the indices of the radicals being multiplied are the same. But always remember to simplify the multiplication results if possible.

Example Multiply the radicals $\sqrt{10} \times \sqrt{5}$.

Solution: According to the product rule: $\sqrt{10} \times \sqrt{5} = \sqrt{10 \times 5} = \sqrt{50}$. Since 50 is not a perfect square, we look for a factor of 50 that is a perfect square to simplify the radical expression: $\sqrt{50} = \sqrt{25 \times 2} = 5\sqrt{2}$.

Example Multiply the following $\sqrt[3]{4} \times \sqrt[3]{4}$.

Solution: When multiplying radicals with the same index, we apply the same product rule for radicals: $\sqrt[3]{4} \times \sqrt[3]{4} = \sqrt[3]{4 \times 4} = \sqrt[3]{16}$. The simplified form of $\sqrt[3]{16}$ is $\sqrt[3]{8 \times 2} = 2\sqrt[3]{2}$.

9.4 Rationalizing Radical Expressions

It is often imperative to rewrite expressions so that square roots are not present in the denominator of fractions. This process, known as rationalizing the denominator.

Key Point

Rationalizing a denominator involves multiplying it by a radical that will eliminate the square root or cube root, leaving only rational numbers. For binomial denominators, we employ the conjugate to eliminate the square roots in the denominator.

The conjugate of a binomial is simply the same two terms with the sign changed in the middle: The conjugate of $a + \sqrt{b}$ would be $a - \sqrt{b}$, and vice versa.

Example Rationalize $\frac{2}{\sqrt{5}}$.

Solution: To rationalize the denominator, we multiply both the numerator and denominator by $\sqrt{5}$: $\frac{2}{\sqrt{5}} \times \frac{\sqrt{5}}{\sqrt{5}} = \frac{2\sqrt{5}}{5}$.

Example Rationalize $\frac{1}{\sqrt{7}-2}$.

Solution: We will multiply the fraction by the conjugate of the denominator $\sqrt{7}+2$: $\frac{1}{\sqrt{7}-2} \times \frac{\sqrt{7}+2}{\sqrt{7}+2} = \frac{\sqrt{7}+2}{7-4} = \frac{\sqrt{7}+2}{3}$.

9.5 Solving Radical Equations

A radical equation is one in which a variable is enclosed within a radical sign. The primary goal when solving such an equation is to first eliminate the radical(s), and then solve the equation that results from this process.

Key Point

To solve radical equations, isolate the radical on one side. Then square or raise to index for both sides of the equation to eliminate the radical. Solve the equation. Always check solutions to avoid extraneous solutions.

Example Solve the radical equation: $\sqrt{x} = 5$

Solution: Square both sides of the equation to eliminate the square root: $(\sqrt{x})^2 = 5^2$. This results in: $x = 25$.

Now, we check this solution by substituting 25 back into the original equation: $\sqrt{25} = 5$. The solution is correct as $5 = 5$.

Example Solve the radical equation: $\sqrt{2x+3} = 4$

Solution: Square both sides to eliminate the square root: $(\sqrt{2x+3})^2 = 4^2$. This simplifies to: $2x+3 = 16$. Subtract 3 from both sides: $2x = 13$. Finally, divide by 2 to isolate x: $x = \frac{13}{2}$.

Check the solution by substituting back into the original equation: $\sqrt{2(\frac{13}{2})+3} = 4$. The solution is correct as $4 = 4$.

9.6 Determining Domain and Range of Radical Functions

The domain of a function includes all possible input values (x-values), while the range of a function includes all possible output values (y-values).

🔔 Key Point

The domain of a radical function with an even root is only those x values that result in a number greater than or equal to zero under the radical. For an odd root, the domain is equal to the domain of the expression under the radical.

🔔 Key Point

The range of a radical function depends on whether the root is even or odd. For a radical function of the form $y = c\sqrt[n]{ax+b} + k$, where n is an even integer, the range depends on the sign of c. If $c > 0$, the range is $y \geq k$. If $c < 0$, the range is $y \leq k$, assuming $ax+b$ is non-negative to ensure the radical is defined. For odd indices, the range is all real numbers, reflecting the unbounded nature of odd root functions over the domain of all real numbers.

📋 Example

Determine the domain and range of the function $f(x) = \sqrt{x}$.

Solution: The function $f(x) = \sqrt{x}$, has a square root (an even root). So, the domain can only include x values that result in a number greater than or equal to zero under the radical. That is, $x \geq 0$.

To find the range, we analyze that this function will always yield real numbers greater than or equal to zero. Hence, the range is $y \geq 0$.

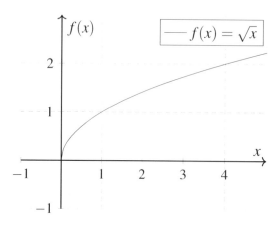

9.7 Simplifying Radicals with Fractional Components

Example Determine the domain and range of the function $f(x) = 3\sqrt{x} - 2$.

Solution: The function $f(x) = 3\sqrt{x} - 2$ also has a square root (an even root) and so, the domain is $x \geq 0$.

To find the range, we multiply real numbers greater than or equal to 0 by 3 and subtract 2. Hence, the range is $y \geq -2$.

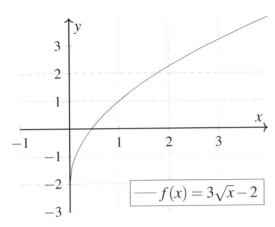

9.7 Simplifying Radicals with Fractional Components

A radical with a fraction radicand is an expression of the form $\sqrt[n]{\frac{a}{b}}$, where a and b are integers and $b \neq 0$. To simplify such an expression, we can use the following steps:

- Rewrite the numerator and denominator of the fraction as the product of the prime factorizations.
- Split the radical into two radicals using the property $\sqrt[n]{\frac{a}{b}} = \frac{\sqrt[n]{a}}{\sqrt[n]{b}}$.
- Simplify each radical separately by finding perfect square or cube factors and taking them out of the radical.

Key Point

To simplify a radical fraction, factorize the numerator and denominator, and apply $\sqrt[n]{\frac{a}{b}} = \frac{\sqrt[n]{a}}{\sqrt[n]{b}}$, where a and b are nonnegative if n is even. Simplify each radical by extracting perfect powers.

Example Simplify the radical expression $\sqrt{\frac{1}{4}}$.

Solution: Since $\sqrt{\frac{1}{4}} = \frac{\sqrt{1^2}}{\sqrt{2^2}}$, we can individually take the square root of the numerator and the denominator. Hence, $\sqrt{\frac{1}{4}} = \frac{1}{2}$.

Example Simplify the radical expression $\sqrt[3]{\frac{8}{27}}$.

Solution: The given expression can be written as $\sqrt[3]{\frac{8}{27}} = \frac{\sqrt[3]{2^3}}{\sqrt[3]{3^3}}$, which allows us to apply the cube root to the numerator and the denominator individually, resulting in $\frac{2}{3}$.

9.8 Practices

Simplify Each Expression:

1) Simplify $\sqrt{72}$.

2) Simplify $3\sqrt{12}$.

3) Simplify $\sqrt{128}$.

4) Simplify $\sqrt{450}$.

5) Simplify $\sqrt{294}$.

Solve:

6) $5\sqrt{7} + 3\sqrt{7}$

7) $4\sqrt{9} - 2\sqrt{9}$

8) $6\sqrt{5} + 2\sqrt{8}$

9) $10\sqrt{3} - 5\sqrt{3}$

10) $3\sqrt{10} + 4\sqrt{6}$

Fill in the Blank:

9.8 Practices

11) Process the multiplying operation: $\sqrt{5} \times \sqrt{7} = $ _____ .

12) Find the simplified form of the multiplication: $\sqrt{24} \times \sqrt{6} = $ _____ .

13) Process the following multiplication: $\sqrt{5} \times \sqrt[3]{125} = $ _____ .

14) Simplify the given radical expression: $\sqrt{9} \times \sqrt{16} = $ _____ .

15) Fill in the simplified form: $\sqrt[4]{16} \times \sqrt[4]{2} = \sqrt[4]{\underline{\hspace{2cm}}}$.

Simplify Each Expression:

16) Rationalize the expression $\frac{3}{\sqrt{2}}$.

17) Rationalize the expression $\frac{4}{\sqrt{7}}$.

18) Rationalize the expression $\frac{5}{\sqrt{6}}$.

19) Rationalize the expression $\frac{1}{2-\sqrt{3}}$.

20) Rationalize the expression $\frac{1}{5-\sqrt{2}}$.

Solve:

21) Solve the equation $\sqrt{3x+7} = 5$.

22) Solve the equation $\sqrt{4x-1} = 7$.

23) Solve the equation $\sqrt{5x+4} = 2$.

24) Solve the equation $\sqrt{2x-5} = 3$.

25) Solve the equation $\sqrt{x+12} = 10$.

Solve:

26) Determine the domain and range of $f(x) = \sqrt{x-3}$.

27) Determine the domain and range of $f(x) = -2\sqrt{x} + 1$.

28) Determine the domain and range of $f(x) = \sqrt{5-x}$.

29) Determine the domain and range of $f(x) = 3\sqrt{x+2} - 1$.

30) Determine the domain and range of $f(x) = -\sqrt{x+4}$.

Simplify Each Expression:

31) Simplify the radical expression $\sqrt{\frac{49}{25}}$.

32) Simplify the radical expression $\sqrt[3]{\frac{125}{64}}$.

33) Simplify the radical expression $\sqrt{\frac{1}{9}}$.

34) Simplify the radical expression $\sqrt[3]{\frac{1}{8}}$.

35) Simplify the radical expression $\sqrt{\frac{36}{16}}$.

Answer Keys

1) $6\sqrt{2}$
2) $6\sqrt{3}$
3) $8\sqrt{2}$
4) $15\sqrt{2}$
5) $7\sqrt{6}$
6) $8\sqrt{7}$
7) $2\sqrt{9}$
8) $6\sqrt{5}+2\sqrt{8}$
9) $5\sqrt{3}$
10) $3\sqrt{10}+4\sqrt{6}$
11) $\sqrt{35}$
12) 12
13) $5\sqrt{5}$
14) 12
15) $2\sqrt[4]{2}$
16) $\frac{3\sqrt{2}}{2}$
17) $\frac{4\sqrt{7}}{7}$
18) $\frac{5\sqrt{6}}{6}$
19) $\frac{2+\sqrt{3}}{1}$
20) $\frac{5+\sqrt{2}}{23}$
21) $x=6$
22) $x=12.5$
23) $x=0$
24) $x=7$
25) $x=88$
26) $D_f : x \geq 3, R_f : y \geq 0$
27) $D_f : x \geq 0, R_f : y \leq 1$
28) $D_f : x \leq 5, R_f : y \geq 0$
29) $D_f : x \geq -2, R_f : y \geq -1$
30) $D_f : x \geq -4, R_f : y \leq 0$
31) $\frac{7}{5}$
32) $\frac{5}{4}$
33) $\frac{1}{3}$
34) $\frac{1}{2}$
35) $\frac{3}{2}$

Answers with Explanation

1) Factor 72 into $2 \times 2 \times 2 \times 3 \times 3$. The pairs of factors (2×2) and (3×3) can be taken out from under the radical, resulting in $2 \times 3 = 6$. Remaining under the radical is 2, hence we get $6\sqrt{2}$.

2) Factor 12 into $2 \times 2 \times 3$. The pair of factors 2×2 can be taken out from under the radical, resulting in 2. This 2 multiplies with the 3 outside the radical to get 6. Remaining under the radical is 3, hence we get $6\sqrt{3}$.

3) Factor 128 into $2 \times 2 \times 2 \times 2 \times 2 \times 2 \times 2$. The pairs of factors can be taken out from under the radical, resulting in $2 \times 2 \times 2 = 8$. Remaining under the radical is 2, hence we get $8\sqrt{2}$.

4) Factor 450 into $2 \times 3 \times 3 \times 5 \times 5$. The pairs of factors (3×3) and (5×5) can be taken out from under the radical, resulting in $3 \times 5 = 15$. Remaining under the radical is 2, hence we get $15\sqrt{2}$.

5) Factor 294 into $2 \times 3 \times 7 \times 7$. The pair of factors (7×7) can be taken out from under the radical, resulting in 7. Remaining under the radical is $(2 \times 3) = 6$, hence we get $7\sqrt{6}$.

6) Both terms are like radicals, so we simply add the coefficients: $5 + 3 = 8$, giving us $8\sqrt{7}$.

7) Both terms are like radicals, so we subtract the coefficients: $4 - 2 = 2$, giving us $2\sqrt{9}$.

8) The radicals are not alike, therefore, we cannot simplify any further.

9) Both terms are like radicals, so we subtract the coefficients: $10 - 5 = 5$, giving us $5\sqrt{3}$.

10) The radicals are not alike, therefore, we can't simplify any further.

11) According to the product rule, multiply the numbers under the radicals to get: $\sqrt{5 \times 7} = \sqrt{35}$.

12) Apply the product rule: $\sqrt{24 \times 6} = \sqrt{144} = \sqrt{12 \times 12} = 12$.

13) $\sqrt[3]{125} = 5$. Therefore: $\sqrt{5} \times \sqrt[3]{125} = 5\sqrt{5}$.

14) $\sqrt{9 \times 16} = \sqrt{144} = 12$. Since 144 is a perfect square, its square root is 12.

9.8 Practices

15) $\sqrt[4]{16 \times 2} = \sqrt[4]{32} = \sqrt[4]{2^4 \times 2} = 2\sqrt[4]{2}$.

16) The idea here is to multiply both the numerator and denominator by $\sqrt{2}$, therefore: $\frac{3}{\sqrt{2}} \times \frac{\sqrt{2}}{\sqrt{2}} = \frac{3\sqrt{2}}{2}$.

17) Proceed by multiplying both the numerator and denominator by $\sqrt{7}$, hence: $\frac{4}{\sqrt{7}} \times \frac{\sqrt{7}}{\sqrt{7}} = \frac{4\sqrt{7}}{7}$.

18) We have to multiply both the numerator and the denominator by $\sqrt{6}$, which provides: $\frac{5}{\sqrt{6}} \times \frac{\sqrt{6}}{\sqrt{6}} = \frac{5\sqrt{6}}{6}$.

19) Multiply the fraction by the conjugate of the denominator, $2 + \sqrt{3}$: $\frac{1}{2-\sqrt{3}} \times \frac{2+\sqrt{3}}{2+\sqrt{3}} = \frac{2+\sqrt{3}}{4-3} = \frac{2+\sqrt{3}}{1}$.

20) Multiply the fraction by the conjugate of the denominator, $5 + \sqrt{2}$: $\frac{1}{5-\sqrt{2}} \times \frac{5+\sqrt{2}}{5+\sqrt{2}} = \frac{5+\sqrt{2}}{25-2} = \frac{5+\sqrt{2}}{23}$.

21) Squaring both sides to eliminate the square root results in: $(3x + 7) = 25$. Subtracting 7 from both sides gives: $3x = 18$. Dividing by 3 to isolate x gives us: $x = 6$

22) Squaring both sides to eliminate the square root gets us: $4x - 1 = 49$. Add 1 to both sides, we get: $4x = 50$. Divide by 4 to isolate x, we get: $x = 12.5$.

23) Squaring both sides gets us: $5x + 4 = 4$. Subtracting 4 from both sides gives us: $5x = 0$. Finally, dividing by 5 to solve for x, we get: $x = 0$.

24) Squaring both sides gets us: $2x - 5 = 9$. Adding 5 to both sides, we get: $2x = 14$. Finally, dividing by 2 to solve for x, we get: $x = 7$.

25) Squaring both sides gets us: $x + 12 = 100$. Subtracting 12, we get: $x = 88$.

26) The function has a square root, meaning x values can only be greater than or equal to zero under the radical. Hence, $x - 3 \geq 0$ leading to $x \geq 3$. The function will always yield real numbers greater than or equal to zero, giving a range of $y \geq 0$.

27) The function has a square root, therefore the domain is $x \geq 0$. Also, because of the -2 coefficient, the function is flipped about the x-axis, and the $+1$ shifts the function up. So, $y \leq 1$.

28) As the function has a square root, x has to be less than or equal to 5 to give positive input under the radical; hence, $5 - x \geq 0$ leading to $x \leq 5$. The function will output real number greater than or equal to zero, which gives a range of $y \geq 0$.

29) The function has a square root, so the domain is $x \geq -2$. Notice the -1 shifts the output down, thus y must be greater than or equal to -1. Therefore, the range is $y \geq -1$.

30) The function has a negation and square root, hence $x+4 \geq 0$ leading to $x \geq -4$ for the domain. The function output is mirrored about the x-axis to give negatives, hence range is $y \leq 0$.

31) The original expression can be rewritten as $\frac{\sqrt{49}}{\sqrt{25}}$, which simplifies to $\frac{7}{5}$.

32) The given expression can be written as $\sqrt[3]{5^3 \div 4^3}$, which allows us to apply the cube root to the numerator and the denominator individually, resulting in $\frac{5}{4}$.

33) Since $\sqrt{\frac{1}{9}} = \sqrt{1^2 \div 3^2}$, we can individually take the square root of the numerator and the denominator. Hence, $\sqrt{\frac{1}{9}} = \frac{1}{3}$.

34) The given expression can be written as $\sqrt[3]{1^3 \div 2^3}$, which allows us to apply the cube root to the numerator and the denominator individually, resulting in $\frac{1}{2}$.

35) The original expression can be rewritten as $\frac{\sqrt{36}}{\sqrt{16}}$, which simplifies to $\frac{3}{2}$.

10. Rational Expressions

10.1 Simplifying Complex Fractions

Complex fractions are fractions where the numerator, denominator, or both contain a fraction. To simplify a complex fraction, you need to carry out the following steps:

- Find the least common denominator (LCD) for all fractions within the complex fraction. Although there may exist multiple LCDs, it is generally the least cumbersome to compute with the smallest among them.
- Substitute this "unified" denominator into each minor fraction in the numerator and denominator separately and simplify these fractions.
- With the simplified form of the complex fraction, carry out the division operation to yield a simplified form of the original complex fraction.

Key Point

To simplify a complex fraction, unify denominators in numerator and denominator by finding the LCD. Simplify numerator and denominator of the complex fraction separately using this LCD. Carry out the division operation in the complex fraction with simplified numerator and denominator.

 Simplify the complex fraction $\frac{\frac{3}{4}-\frac{1}{2}}{\frac{1}{3}+\frac{1}{2}}$.

Solution: We start by finding the LCD. The LCD of 4, 2, 3, 2 is 12. Next, we unify our denominators:

$$\frac{\frac{3\times 3}{3\times 4} - \frac{1\times 6}{2\times 6}}{\frac{1\times 4}{3\times 4} + \frac{1\times 6}{2\times 6}} = \frac{\frac{9}{12} - \frac{6}{12}}{\frac{4}{12} + \frac{6}{12}}.$$

We simplify the numerator and the denominator separately:

$$\frac{\frac{9}{12} - \frac{6}{12}}{\frac{4}{12} + \frac{6}{12}} = \frac{\frac{3}{12}}{\frac{10}{12}}.$$

Finally, simplifying the fraction:

$$\frac{\frac{3}{12}}{\frac{10}{12}} = \frac{3}{12} \times \frac{12}{10} = \frac{3}{10}.$$

Therefore, $\frac{\frac{3}{4}-\frac{1}{2}}{\frac{1}{3}+\frac{1}{2}} = \frac{3}{10}$.

 Example Simplify the complex fraction $\frac{\frac{2}{3}+\frac{1}{2}}{\frac{5}{6}-\frac{1}{3}}$.

Solution: The LCD of 3, 2, 6, 3 is 6. To unify the denominators, we have:

$$\frac{\frac{2\times 2}{3\times 2} + \frac{1\times 3}{2\times 3}}{\frac{5\times 1}{6\times 1} - \frac{1\times 2}{3\times 2}} = \frac{\frac{4}{6} + \frac{3}{6}}{\frac{5}{6} - \frac{2}{6}}.$$

Simplifying the numerator and the denominator separately, we have:

$$\frac{\frac{4}{6} + \frac{3}{6}}{\frac{5}{6} - \frac{2}{6}} = \frac{\frac{7}{6}}{\frac{3}{6}}.$$

Finally, dividing the fraction simplifies to:

$$\frac{\frac{7}{6}}{\frac{3}{6}} = \frac{7}{6} \times \frac{6}{3} = \frac{7}{3}.$$

Therefore, $\frac{\frac{2}{3}+\frac{1}{2}}{\frac{5}{6}-\frac{1}{3}} = \frac{7}{3}$.

10.2 Graphing Rational Functions

A rational function is a function of the form $f(x) = \frac{p(x)}{q(x)}$ where $p(x)$ and $q(x)$ are polynomial functions and $q(x) \neq 0$. When graphing such functions, we need to investigate four key aspects: the domain, the vertical and horizontal asymptotes, and the intercepts.

> **Key Point**
>
> The domain of a rational function consists of all real numbers except for the zeroes of the denominator $q(x)$.

Zeroes of the denominator cause the function to be undefined and they correspond to the vertical asymptotes on the graph.

> **Key Point**
>
> A vertical asymptote is a vertical line $x = a$, where a is a root of the denominator $q(x)$. The function tends towards $\pm\infty$ when x approaches a.

The function has a horizontal asymptote if the degree of the polynomial in the numerator is less than or equal to the degree of the polynomial in the denominator.

> **Key Point**
>
> A horizontal asymptote is a horizontal line $y = b$, where $b = \frac{p(n)}{q(n)}$ for large n if the degree of $p(x)$ is less than or equal to the degree of $q(x)$.

Finding the intercepts helps us to anchor the graph in specific points.

> **Key Point**
>
> The x-intercepts are the roots of the numerator $p(x)$ and the y-intercept is found by substituting $x = 0$ into the function.

Example Plot the rational function $f(x) = \frac{x}{x-2}$.

Solution: The domain of the function is all real numbers except $x = 2$. So, we have a vertical asymptote at $x = 2$.

The degree of the numerator is equal to the degree of the denominator, hence, the horizontal asymptote is given by the ratio of the leading coefficients, which is $y = 1$.

The x-intercept is $x = 0$ and by substituting $x = 0$ into the function we get the y-intercept $y = 0$.

Thus, by plotting the points around the intercepts and the asymptotes, we get the graph of the function:

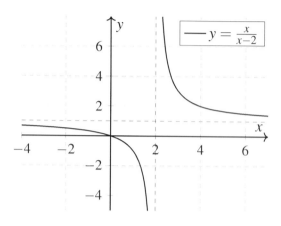

 **Example** Plot the rational function $f(x) = \frac{x+2}{x^2-4}$.

Solution: In fact $f(x) = \frac{1}{x-2}$ for $x \neq 2, -2$. So the function f is undefined at -2 but has not vertical asymptote at this point. The denominator is 0 at $x = 2$, so we have one vertical asymptote at this value.

The degree of the numerator is less than the degree of the denominator, thus our horizontal asymptote is $y = 0$.

The function does not have an x-intercept because there is no value of x for which $f(x) = 0$. By substituting $x = 0$ into the function, we find the y-intercept $y = -0.5$, indicating the point where the graph crosses the y-axis.

By plotting the intercept and the asymptotes, we can sketch the graph of the function:

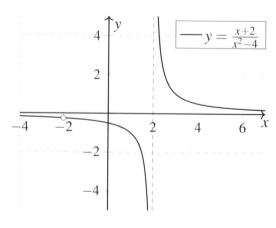

10.3 Performing Addition and Subtraction of Rational Expressions

Similar to regular fractions, addition or subtraction of rational expressions requires a common denominator. Therefore, the first step in either operation is to determine the Least Common Denominator (LCD). It is the least common multiple (LCM) of the denominators of the rational expressions.

Once the LCD has been established, each rational expression is rewritten as an equivalent expression with the LCD as the denominator. Now, the numerators can be added or subtracted.

> **Key Point**
>
> The Least Common Denominator (LCD) of rational expressions is found by determining the Least Common Multiple (LCM) of the polynomials in the denominator.

> **Key Point**
>
> Before adding or subtracting, rational expressions need to have a common denominator. Each expression is rewritten with the LCD as the denominator. Then, add or subtract the numerators and simplify as needed.

Example Add the rational expressions $\frac{3x}{x^2-4}$ and $\frac{2x-6}{x-2}$.

Solution: First, notice that $x^2 - 4 = (x-2)(x+2)$. Thus, the LCM (LCD) of (x^2-4) and $(x-2)$ is $(x-2)(x+2)$.

The expression $\frac{2x-6}{x-2}$ can be rewritten as:

$$\frac{2x-6}{x-2} \times \frac{x+2}{x+2} = \frac{2x^2 - 2x - 12}{(x-2)(x+2)}.$$

The result is:

$$\frac{3x}{x^2-4} + \frac{2x^2 - 2x - 12}{(x-2)(x+2)} = \frac{2x^2 + x - 12}{x^2 - 4}.$$

 **Example** Subtract the rational expression $\frac{3x}{x^2-4}$ from $\frac{2x-6}{x-2}$.

Solution: With the same procedure as in the last example, we find the LCD and rewrite both expressions with the LCD as the denominator. The result is:

$$\frac{2x^2 - 2x - 12}{x^2 - 4} - \frac{3x}{x^2 - 4} = \frac{2x^2 - 5x - 12}{x^2 - 4}.$$

10.4 Performing Multiplication of Rational Expressions

Multiplying rational expressions is similar to multiplying numerical fractions. However, as these expressions involve variables, a few extra steps are involved.

The process begins with factorization which simplifies the expressions and makes it easier to identify. Now, cancel out common factors between numerators and denominators, leading to the simplified result.

> The process of multiplying rational expressions involves factoring, identifying common factors in the numerators and denominators, and cancelling them out.

 **Example** Multiply the rational expressions $\frac{3x}{x-4}$ and $\frac{x+2}{3}$.

Solution: First, we multiply the numerators together and the denominators together, as we do with numerical fractions:

$$\frac{3x}{x-4} \times \frac{x+2}{3} = \frac{3x(x+2)}{(x-4) \times 3}.$$

Next, we observe that both the numerator and denominator can be divided by 3:

$$\frac{3x(x+2)}{(x-4)3} = \frac{x^2 + 2x}{x-4}.$$

 **Example** Multiply the rational expressions $\frac{3x^2 - 12x}{x^2 - 4}$ and $\frac{x^2 - 4}{9x - 36}$.

Solution: First, we factorize the expressions completely:

$$\frac{3x^2 - 12x}{x^2 - 4} = \frac{3x(x-4)}{(x-2)(x+2)} \quad \text{and} \quad \frac{x^2 - 4}{9x - 36} = \frac{(x-2)(x+2)}{9(x-4)}.$$

Next, we multiply like we do numerical fractions, but recall the common factors:

$$\frac{3x(x-4)}{(x-2)(x+2)} \times \frac{(x-2)(x+2)}{9(x-4)} = \frac{3x}{9}.$$

Finally, we simplify the rational expression: $\frac{3x}{9} = \frac{x}{3}$.

10.5 Performing Division of Rational Expressions

The division of rational expressions is not different from the division of regular numbers. The process of dividing fractions involves flipping the divisor and then changing the division operation to multiplication.

When dealing with rational expressions, the same steps are followed. The only addition is that you may have to simplify the expressions first before performing the division, for simplicity. The overall process can be condensed into three steps:

1. Factorize all the expressions completely.

2. Change the division operation to multiplication and flip the divisor.

3. Perform the multiplication and cancellation, just like the multiplication of rational expressions.

Key Point

The division of rational expressions requires factorization of expressions, changing division to multiplication by flipping the divisor, and cancelling out common factors.

 Divide the rational expressions $\frac{x^2-4}{3x}$ by $\frac{x-2}{3}$.

Solution: First, factorize expressions: $\frac{x^2-4}{3x}$ becomes $\frac{(x-2)(x+2)}{3x}$.

Next, change the division to multiplication and flip the divisor:

$$\frac{x^2-4}{3x} \div \frac{x-2}{3} = \frac{(x-2)(x+2)}{3x} \times \frac{3}{x-2}.$$

After that, perform multiplication and cancellation, with common factors in numerators and denominators cancelling each other out:

$$\frac{(x-2)(x+2)}{3x} \times \frac{3}{x-2} = \frac{(x+2)}{x}.$$

 **Example** Divide the rational expressions $\frac{3x^2-12x}{x^2-4}$ by $\frac{9x-36}{x^2-4}$.

Solution: Firstly, factorize expressions completely:

$$\frac{3x^2-12x}{x^2-4} = \frac{3x(x-4)}{(x-2)(x+2)} \quad \text{and} \quad \frac{9x-36}{x^2-4} = \frac{9(x-4)}{(x+2)(x-2)}.$$

Change the division to multiplication and flip divisor:

$$\frac{3x^2-12x}{x^2-4} \div \frac{9x-36}{x^2-4} = \frac{3x(x-4)}{(x-2)(x+2)} \times \frac{(x+2)(x-2)}{9(x-4)}.$$

Finish with multiplication and cancellation:

$$\frac{3x(x-4)}{(x-2)(x+2)} \times \frac{(x+2)(x-2)}{9(x-4)} = \frac{x}{3}.$$

10.6 Evaluating Integers with Rational Exponents

The process of evaluating integers with rational exponents involves understanding and applying the concept of roots and powers. A rational exponent is just an alternative way of expressing roots and fractions.

The general form for expressing a rational exponent is $a^{\frac{m}{n}}$. This form should be understood as the nth root of a^m, where a is the base, m is the numerator, the power, and n is the denominator, the root.

Key Point

> An integer with a rational exponent $a^{\frac{m}{n}}$ can be evaluated by raising a to the power m and taking the nth root of the result, or by taking the nth root of a and raising the result to the power m.

 Example Evaluate the integer with a rational exponent, $8^{\frac{2}{3}}$.

Solution: We know that $8^{\frac{2}{3}} = (8^2)^{\frac{1}{3}}$ or $(8^{\frac{1}{3}})^2$. Using the first order of operations, we get $8^{\frac{2}{3}} = (64)^{\frac{1}{3}}$ which simplifies to 4.

Alternatively, taking the cube root of 8 first $(8^{\frac{1}{3}}) = 2$ and squaring it to the 2nd power gives us $2^2 = 4$. So, $8^{\frac{2}{3}} = 4$.

10.7 Practices

 Example Evaluate the integer with a rational exponent, $16^{\frac{3}{4}}$.

Solution: We can write $16^{\frac{3}{4}} = (16^3)^{\frac{1}{4}}$ or $(16^{\frac{1}{4}})^3$. First, we can find the 4th root of 16, $16^{\frac{1}{4}} = 2$. Next, we raise 2 to the power of 3, $2^3 = 8$. Therefore, $16^{\frac{3}{4}} = 8$.

10.7 Practices

Simplify Each Expression:

1) Simplify the complex fraction $\dfrac{\frac{2}{3} - \frac{1}{4}}{\frac{5}{6} + \frac{1}{2}}$.

2) Simplify the complex fraction $\dfrac{\frac{3}{2} + \frac{1}{5}}{\frac{4}{3} - \frac{1}{4}}$.

3) Simplify the complex fraction $\dfrac{\frac{1}{4} - \frac{3}{5}}{\frac{2}{3} + \frac{1}{4}}$.

4) Simplify the complex fraction $\dfrac{\frac{2}{5} + \frac{3}{4}}{\frac{1}{6} - \frac{1}{3}}$.

5) Simplify the complex fraction $\dfrac{\frac{5}{6} - \frac{2}{3}}{\frac{3}{4} + \frac{1}{2}}$.

Identify the Key Characteristics:

6) For the function $f(x) = \frac{x-1}{x+2}$, find the domain, vertical and horizontal asymptotes, and the x- and y-intercepts.

7) For the function $f(x) = \frac{x^2-1}{x-2}$, find the domain, vertical and horizontal asymptotes, and the x- and y-intercepts.

8) For the function $f(x) = \frac{3x+4}{x-2}$, find the domain, vertical and horizontal asymptotes, and the x- and y-intercepts.

9) For the function $f(x) = \frac{x+2}{2x+1}$, find the domain, vertical and horizontal asymptotes, and the x- and y-intercepts.

10) For the function $f(x) = \frac{x^2+1}{x+3}$, find the domain, vertical and horizontal asymptotes, and the x- and y-intercepts.

 Solve:

11) Add the rational expressions $\frac{7x}{x^2-9}$ and $\frac{5x+15}{x+3}$.

12) Subtract the rational expression $\frac{4x-16}{x-4}$ from $\frac{2x}{x^2-16}$.

13) Simplify $\frac{x^2-4}{x^2-1} - \frac{3x+3}{x-1}$.

14) Add the rational expressions $\frac{x}{x^2-4}$ and $\frac{-3x-12}{x+4}$.

15) Subtract the fraction $\frac{3x-8}{x-2}$ from $\frac{6x}{x^2-4}$.

Simplify:

16) Multiply the rational expressions $\frac{2x}{x-3}$ and $\frac{x+4}{5}$.

17) Multiply the rational expressions $\frac{2x}{x^2-9}$ and $\frac{x^2-1}{2x}$.

18) Multiply these rational expressions $\frac{x^2}{x+5}$ and $\frac{x+5}{x^2}$.

19) Multiply the rational expressions $\frac{4x}{2x+2}$ and $\frac{2x+2}{4x}$.

20) Multiply these rational expressions $\frac{3x^2}{x-2}$ and $\frac{2-x}{3x}$.

Solve:

21) $\frac{4x-20}{x-5} \div \frac{2x}{5} =$

22) $\frac{2x^2-18}{x+3} \div \frac{2x-6}{x} =$

23) $\frac{3x^3-3x^2}{x-2} \div \frac{3x^2-6x}{x} =$

24) $\frac{5x^2-20x}{x-4} \div \frac{5x-10}{2} =$

25) $\frac{x^3-x^2}{x-1} \div \frac{x^2-x}{1} =$

Equation Solving:

26) Solve for x: $x^{\frac{2}{3}} = 8$

10.7 Practices

27) Solve for x: $x^{\frac{3}{2}} = 27$

28) Solve for x: $x^{\frac{1}{3}} = 4$

29) Solve for x: $(x^{\frac{1}{4}})^2 = 4$

30) Solve for x: $(x^{\frac{1}{2}})^3 = 8$

Answer Keys

1) $\frac{5}{16}$

2) $\frac{102}{65}$

3) $-\frac{21}{55}$

4) $-\frac{69}{10}$

5) $\frac{2}{15}$

6) $D: x \neq -2, VA: x = -2, HA: y = 1$, x-intercept: $(1,0)$, y-intercept: $(0, -\frac{1}{2})$

7) $D: x \neq 2, VA: x = 2, HA:$ No asymptote, x-intercepts: $(1,0), (-1,0)$, y-intercept: $(0, \frac{1}{2})$.

8) $D: x \neq 2, VA: x = 2, HA: y = 3$, x-intercept: $(-\frac{4}{3}, 0)$, y-intercept: $(0, -2)$

9) $D: x \neq -\frac{1}{2}, VA: x = -\frac{1}{2}, HA: y = \frac{1}{2}$, x-intercept: $(-2, 0)$, y-intercept: $(0, 2)$

10) $D: x \neq -3, VA: x = -3, HA:$ No asymptote, x-intercept: No intercepts, y-intercept: $(0, \frac{1}{3})$

11) $\frac{5x^2+7x-45}{x^2-9}$

12) $\frac{-4x^2+2x+64}{x^2-16}$

13) $\frac{-2x^2-6x-7}{x^2-1}$

14) $\frac{-3x^2+x+12}{x^2-4}$

15) $\frac{-3x^2+8x+16}{x^2-4}$

16) $\frac{2x(x+4)}{5(x-3)} = \frac{2x^2+8x}{5x-15}$

17) $\frac{x^2-1}{x^2-9}$

18) 1

19) 1

20) $-x$

21) $\frac{10}{x}$

22) x

23) $\frac{x^3-x^2}{(x-2)^2}$

24) $\frac{2x}{x-2}$

25) $\frac{x}{x-1}$

26) $x = \pm 16\sqrt{2}$

27) $x = 9$

28) $x = 64$

29) $x = 16$

30) $x = 4$

10.7 Practices

Answers with Explanation

1) The LCD of 3, 4, 6, 2 is 12. Unifying the denominators: $\frac{\frac{2\times 4}{3\times 4}-\frac{1\times 3}{4\times 3}}{\frac{5\times 2}{6\times 2}+\frac{1\times 6}{2\times 6}}=\frac{\frac{8}{12}-\frac{3}{12}}{\frac{10}{12}+\frac{6}{12}}$. Simplifying the numerator and the denominator separately: $\frac{\frac{8}{12}-\frac{3}{12}}{\frac{10}{12}+\frac{6}{12}}=\frac{\frac{5}{12}}{\frac{16}{12}}$. Simplifying the fraction: $\frac{\frac{5}{12}}{\frac{16}{12}}=\frac{5}{12}\times\frac{12}{16}=\frac{5}{16}$.

2) The LCD of 2, 5, 3, 4 is 60. Unifying the denominators: $\frac{\frac{3\times 30}{2\times 30}+\frac{1\times 12}{5\times 12}}{\frac{4\times 20}{3\times 20}-\frac{1\times 15}{4\times 15}}=\frac{\frac{90}{60}+\frac{12}{60}}{\frac{80}{60}-\frac{15}{60}}$. Simplifying the numerator and the denominator separately: $\frac{\frac{90}{60}+\frac{12}{60}}{\frac{80}{60}-\frac{15}{60}}=\frac{\frac{102}{60}}{\frac{65}{60}}$. Simplifying the fraction: $\frac{\frac{102}{60}}{\frac{65}{60}}=\frac{102}{60}\times\frac{60}{65}=\frac{102}{65}$.

3) The LCD of 4, 5, 3, 4 is 60. Unifying the denominators: $\frac{\frac{1\times 15}{4\times 15}-\frac{3\times 12}{5\times 12}}{\frac{2\times 20}{3\times 20}+\frac{1\times 15}{4\times 15}}=\frac{\frac{15}{60}-\frac{36}{60}}{\frac{40}{60}+\frac{15}{60}}$. Simplifying the numerator and the denominator separately: $\frac{\frac{15}{60}-\frac{36}{60}}{\frac{40}{60}+\frac{15}{60}}=\frac{\frac{-21}{60}}{\frac{55}{60}}$. Simplifying the fraction: $\frac{\frac{-21}{60}}{\frac{55}{60}}=\frac{-21}{60}\times\frac{60}{55}=-\frac{21}{55}$.

4) The LCD of 5, 4, 6, 3 is 60. Unifying the denominators: $\frac{\frac{2\times 12}{5\times 12}+\frac{3\times 15}{4\times 15}}{\frac{1\times 10}{6\times 10}-\frac{1\times 20}{3\times 20}}=\frac{\frac{24}{60}+\frac{45}{60}}{\frac{10}{60}-\frac{20}{60}}$. Simplifying the numerator and the denominator separately: $\frac{\frac{24}{60}+\frac{45}{60}}{\frac{10}{60}-\frac{20}{60}}=\frac{\frac{69}{60}}{\frac{-10}{60}}$. Simplifying the fraction: $\frac{\frac{69}{60}}{\frac{-10}{60}}=\frac{69}{60}\times\frac{-60}{10}=-\frac{69}{10}$.

5) The LCD of 6, 3, 4, 2 is 12. Unifying the denominators: $\frac{\frac{5\times 2}{6\times 2}-\frac{2\times 4}{3\times 4}}{\frac{3\times 3}{4\times 3}+\frac{1\times 6}{2\times 6}}=\frac{\frac{10}{12}-\frac{8}{12}}{\frac{9}{12}+\frac{6}{12}}$. Simplifying the numerator and the denominator separately: $\frac{\frac{10}{12}-\frac{8}{12}}{\frac{9}{12}+\frac{6}{12}}=\frac{\frac{2}{12}}{\frac{15}{12}}$. Simplifying the fraction: $\frac{\frac{2}{12}}{\frac{15}{12}}=\frac{2}{12}\times\frac{12}{15}=\frac{2}{15}$.

6) The function is undefined at $x=-2$, hence its domain is all real numbers except -2. A vertical asymptote is at $x=-2$ because this is where the denominator is 0. The degrees of the numerator and the denominator are equal, so the horizontal asymptote is the ratio of the leading coefficients, $y=1$. The x-intercept is found by solving $x-1=0$ which gives $x=1$. The y-intercept is found by substituting $x=0$ in the function which gives $y=-\frac{1}{2}$.

7) The function is undefined at $x=2$, hence its domain is all real numbers except 2. A vertical asymptote is at $x=2$ because this is where the denominator is 0. The degree of the numerator is greater than the denominator, so there is no horizontal asymptote. The x-intercepts are found by solving $x^2-1=0$ which gives $x=\pm 1$. The y-intercept is found by substituting $x=0$ in the function which gives $y=\frac{1}{2}$.

8) The function is undefined at $x=2$, so the domain is all real numbers except 2. The vertical asymptote is at $x = 2$ which corresponds to a zero of the denominator. The degrees of the numerator and denominator are the same, so the horizontal asymptote is at $y = 3$, which is the ratio of leading coefficients. The x-intercept is found by setting the numerator equal to zero, giving $x = -\frac{4}{3}$. The y-intercept is found by substituting $x = 0$ in the function which gives $y = -2$.

9) The function is undefined at $x = -\frac{1}{2}$, hence its domain is all real numbers except $-\frac{1}{2}$. The vertical asymptote is at $x = -\frac{1}{2}$ which corresponds to a zero of the denominator. The degrees of the numerator and denominator are the same, so there is a horizontal asymptote at $y = \frac{1}{2}$, which is the ratio of the leading coefficients. The x-intercept is found by setting the numerator equal to zero, that gives $x = -2$. The y-intercept is found by substituting $x = 0$ in the function which gives $y = 2$.

10) The function is undefined at $x = -3$, hence its domain is all real numbers except -3. A vertical asymptote is at $x = -3$ which corresponds to a zero of the denominator. The degree of the numerator is greater than the denominator, hence there is no horizontal asymptote. A real root does not exist for $x^2 + 1 = 0$, hence there are no x-intercepts. The y-intercept is found by substituting $x = 0$ in the function which gives $y = \frac{1}{3}$.

11) First, factor the denominators. The first denominator, $x^2 - 9$, factors as $(x+3)(x-3)$. The least common denominator (LCD) is $x^2 - 9$.

To add, express both fractions with the LCD. The second expression becomes $\frac{(5x+15)(x-3)}{x^2-9}$ after multiplying the numerator and denominator by $(x-3)$, simplifying to $\frac{5x^2-45}{x^2-9}$.

Adding the fractions gives $\frac{7x+5x^2-45}{x^2-9} = \frac{5x^2+7x-45}{x^2-9}$, the simplified sum.

12) The LCD of the denominators is $x^2 - 16$. Rewrite the second term with the LCD then subtract to obtain the solution.

13) First, identify the LCD, which is $x^2 - 1$. Rewrite the second fraction $\frac{3x+3}{x-1}$ with the LCD by multiplying the numerator and denominator by $x+1$. This step yields $\frac{3x^2+6x+3}{x^2-1}$. Then, subtract this expression from $\frac{x^2-4}{x^2-1}$. Simplify the resulting expression by combining like terms in the numerator to get $\frac{-2x^2-6x-7}{x^2-1}$.

14) The LCD of the denominators is $x^2 - 4$. By rewriting the terms with the LCD and adding, we obtain the solution.

15) The common denominator of the terms is $x^2 - 4$. Alter the terms accordingly and subtract to find the solution.

10.7 Practices

16) We first multiply the numerators i.e., $2x$ and $x+4$ and denominators i.e., $x-3$ and 5, which gives us $\frac{2x^2+8x}{5x-15}$.

17) We simplify the rational expressions by cancelling out the common factors from both the numerator and denominator which is $2x$, leaving only $\frac{x^2-1}{x^2-9}$.

18) Every term in the numerator can be cancelled out by a matching term in the denominator. Thus the simplified answer is 1.

19) Every term in the numerator can be cancelled out by a matching term in the denominator. Thus the simplified answer is 1.

20) We simplify the rational expressions by cancelling ou the common terms from both the numerator and denominator, which gives us $-x$.

21) The division of these two rational expressions after simplification and cancellation gives $\frac{10}{x}$.

22) After factorization, simplification, and cancellation of common factors, the final answer is x.

23) Simplifying this division of rational expressions gives $\frac{x^3-x^2}{(x-2)^2}$.

24) The result of this division of rational expressions is $\frac{2x}{x-2}$.

25) The division result of these rational expressions simplifies to $\frac{x}{x-1}$.

26) By cubing both sides, we obtain $x^2 = 2^9$. Taking the square root of both sides then gives $x = \pm 16\sqrt{2}$.

27) By squaring both sides, we get $x^3 = 9^3$. Therefore $x = 9$.

28) By cubing both sides, we get $x = 64$.

29) Taking the square root on both sides, we get $x^{\frac{1}{4}} = 2$, taking the 4th power again we get $x = 16$.

30) Taking the cube root on both sides, we get $x^{\frac{1}{2}} = 2$, squaring both sides we get $x = 4$.

11. Statistics and Probabilities

11.1 Calculating Mean, Median, Mode, and Range

The **mean** of a data set is the sum of all the data divided by the count of the data. We often refer to the mean as the "average". If you have a data set $\{x_1, x_2, ..., x_n\}$, you calculate the mean by using the formula:

$$\text{Mean} = \frac{1}{n}\sum_{i=1}^{n} x_i.$$

The sigma sign is used to represent the sum.

The **median** of a data set is the middle number when the data are arranged in ascending (or descending) order. If the data set has an odd number of observations, the median is the middle number. If the data set has an even number of observations, the median is the average of the two middle numbers.

The **mode** of a data set is the number or numbers that occur most frequently. A data set may have one mode, more than one mode, or no mode at all.

Finally, the **range** of a data set is the difference between the highest and lowest values in the data set.

Key Point

> The **mean** is the sum of data divided by the total number of data. The **median** is the middle value in ordered data. The **mode** is the most frequent value(s). The **range** is the difference between the highest and lowest values.

Example Calculate the mean, median, mode, and range of the following series: 5, 2, 9, 3, 5, 7, 5.

Solution: To calculate the **mean**, first sum all of the numbers together $5+2+9+3+5+7+5 = 36$. There are seven numbers in this series, so you would divide the sum by seven. Doing this gives us: $\frac{36}{7} \approx 5.14$.

To find the **median**, we need to arrange the numbers in ascending order, 2, 3, 5, 5, 5, 7, 9. Since there are seven numbers, the median is the number in the middle (4th position here), which is 5.

The number that appears most frequently is the **mode**. In this series, the mode is 5, because it appears three times.

To find the **range**, subtract the smallest number from the largest. Therefore, $9-2=7$.

So, the mean is 5.14, the median is 5, the mode is 5 and the range is 7.

 Example Calculate the mean, median, mode, and range of the following series: 6, 8, 10, 8, 14, 16.

Solution: **Mean:** Calculate the sum of the numbers: $6+8+10+8+14+16 = 62$ and divide by the total number of values, 6, to get $\frac{62}{6} \approx 10.33$.

Median: Organize the numbers in ascending order: 6, 8, 8, 10, 14, 16. We have an even number of values, so the median is the mean of the two middle numbers: $8+10 = 18$ and $\frac{18}{2} = 9$.

Mode: The number that appears the most frequently is 8, which happens to be appearing twice.

Range: Subtract the smallest number from the largest to get $16-6 = 10$.

In conclusion, the mean is 10.33, the median is 9, the mode is 8 and the range is 10.

11.2 Creating a Pie Graph

A pie graph, also known as a circle graph, is a circular graph that is divided into slices to represent numerical proportions. Each slice of the pie corresponds to a particular category. The greater the sector area or the angle at the center is, the larger the data it represents.

A pie graph is used to visually represent percentages or proportional data and are divided into sectors, where each sector represents a proportion of the total.

One of the main advantages of a pie graph over other graphs is that it clearly shows the relative size of data, making it easier to understand complex data sets.

To draw a pie graph, follow the given steps:

1. First, calculate the total sum of the data.

2. Calculate the angle that each piece of data should occupy in the pie graph.

As the pie graph represents 100% of the data (which corresponds a full circle of 360°), each data item can be calculated as:
$$\text{angle} = \frac{\text{data value}}{\text{total data sum}} \times 360°.$$

3. Once all angles are calculated, draw a circle and plot each data item as per its calculated angle in the pie graph.

4. Label the graph with data details.

Remember, the sum of all the pie slices should always be 360°

Key Point

To create a pie graph, calculate the total of all data, then determine each slice's angle using $\frac{\text{data value}}{\text{total data sum}} \times 360°$. Draw a circle, plot each slice by its angle, and label accordingly.

Example

Create a pie graph for the categories of a college fund: Tuition: $12000, Books: $800, and Transportation: $700.

Solution: The total sum of the fund is $12000 (Tuition) + $800 (Books) + $700 (Transportation) = $13500.

Calculate the angle each category occupies in the pie graph.

For Tuition, the angle is: $\frac{12000}{13500} \times 360° \approx 320°$.

For Books, the angle is: $\frac{800}{13500} \times 360° \approx 21°$.

For Transportation, the angle is: $\frac{700}{13500} \times 360° \approx 19°$.

The sum of all angles is $320° + 21° + 19° = 360°$.

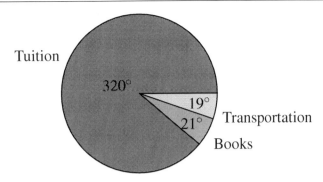

Example The student council conducted a survey to find the favorite sport of the students in a school. The results were:

Soccer: 45 students, Basketball: 30 students, and Volleyball: 25 students.

Create a pie graph for the survey results.

Solution: The total number of students who participated in the survey, which is 45 (Soccer) + 30 (Basketball) + 25 (Volleyball) = 100.

Calculate the angle each sport occupies in the pie graph.

For Soccer, the angle is: $\frac{45}{100} \times 360 = 162°$.

For Basketball, the angle is: $\frac{30}{100} \times 360 = 108°$.

For Volleyball, the angle is: $\frac{25}{100} \times 360 = 90°$.

The sum of all angles is $162° + 108° + 90° = 360°$.

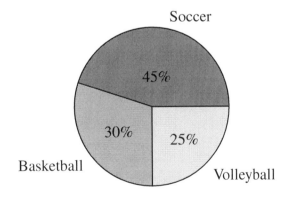

11.3 Analyzing Scatter Plots

A scatter plot, also known as scatter graph or scatter chart, is a plot that uses Cartesian coordinates to represent the relationship between two variables. Remember that a variable is a characteristic that can be measured and that can assume different values. Height, age, income, grades obtained at school are all examples of variables.

The relationship can be positive (both variables increase), negative (one variable decreases as the other increases), or no relationship.

Each dot in a scatter plot corresponds to an observation. To draw a scatter plot, use the given steps:

1. Choose a set of data pairs.

2. Draw an x-y coordinate plane on a graph.

3. Plot the points corresponding to the data pairs on the graph.

Key Point

In scatter plots, a positive trend occurs when y increases with x, and a negative trend occurs when y decreases as x increases.

Example

A student noted the number of hours he studied and his test scores for 5 tests. The data pairs are as follows: (2, 70), (3, 80), (4, 85), (5, 90), (6, 95). Plot these on a scatter plot.

Solution: To plot the scatter plot, first draw an x-y coordinate plane. The x-axis can represent the hours studied and the y-axis can represent the test scores. Now plot the data pairs on the graph. The first pair (2, 70) is plotted as a point 2 units along the x-axis (hours) and 70 units up the y-axis (score). Similarly, plot the other points.

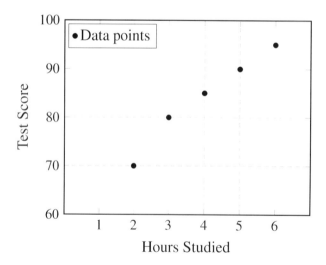

From this scatter plot, we can see a positive trend that as the number of hours studying increases, the test score also increases.

11.4 Calculating and Interpreting Correlation Coefficients

A correlation coefficient is a statistical measure that calculates the strength of the relationship between the relative movements of two variables. The values range between -1 and 1. A positive correlation indicates that the variables move in the same direction, while a negative correlation shows they move in opposite directions.

Key Point

A correlation coefficient quantifies the degree to which two variables are related. A value of $+1$ implies a perfect positive correlation, while -1 implies a perfect negative correlation. A value of 0 implies no correlation.

A common type of correlation coefficient is the sample Pearson correlation coefficient:

Key Point

The sample Pearson correlation coefficient (r) measures the linear relationship between two variables, calculated by

$$r = \frac{1}{n-1} \sum_{i=1}^{n} \left(\frac{x_i - \bar{x}}{s_x}\right)\left(\frac{y_i - \bar{y}}{s_y}\right).$$

Where x and y are the variables we are considering, $\bar{x}$ and $\bar{y}$ are their means, and s_x and s_y are their sample standard deviations.

standard deviation, is a statistical measure which quantifies data dispersion around the mean. A smaller standard deviation suggests data is closely clustered around the mean, while a larger one indicates wider variability. The sample standard deviation formula is:

$$s_x = \sqrt{\frac{1}{n-1} \sum_{i=1}^{n} (x_i - \bar{x})^2}.$$

Example

Find the sample Pearson correlation coefficient between x and y for this data $(1,2), (2,3), (3,2), (4,4), (5,5)$.

Solution:

First we need to calculate the means: $\bar{x} = 3$ and $\bar{y} = 3.2$.

Then calculate the sample standard deviations: $s_x \approx 1.58$ and $s_y \approx 1.3$.

Finally, substitute the values into the Pearson correlation coefficient formula: $r \approx 0.85$.

So, x and y have a strong positive correlation. The following scatter plot shows this positive correlation.

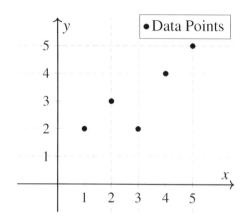

Example Given two sets of numbers $A = \{1,2,3,4,5\}$ and $B = \{5,4,3,2,1\}$, calculate the correlation coefficient between them.

Solution:

The means are $\overline{A} = 3$ and $\overline{B} = 3$.

The standard deviations are $s_A \approx 1.58$ and $s_B \approx 1.58$.

Substitute into the Pearson correlation coefficient formula: $r = \approx -1$.

Hence, A and B have a perfect negative correlation. See the following scatter plot of variables A and B.

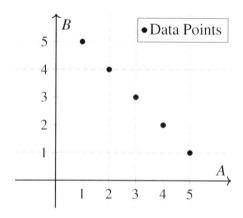

11.5 Formulating the Equation of a Regression Line

A regression line can be thought of as a "best fit" line through a scatter plot of data points. Conceptually, the line minimizes the distance from each data point to the line itself. This is done by minimizing the sum of

11.5 Formulating the Equation of a Regression Line

the squares of the vertical distances from each point to the line; thus, it is often referred to as least squares regression.

The equation of a regression line is given by $y = mx + b$. The slope m and y-intercept b are calculated as follows:

$$m = r \times \frac{s_y}{s_x} \quad \text{and} \quad b = \bar{y} - m\bar{x},$$

where:
- r is the correlation coefficient.
- s_x and s_y are the standard deviations of x and y, respectively.
- $\bar{x}$ and $\bar{y}$ are the means of x and y, respectively.

Key Point

A regression line minimizes the sum of squares of vertical distances from each data point to the line. Its equation, $y = mx + b$, where $m = r \times \frac{s_y}{s_x}$ and $b = \bar{y} - m\bar{x}$, depends on the correlation coefficient r, standard deviations s_x, s_y, and means $\bar{x}, \bar{y}$.

Example

Given the sets of values $X = \{1, 2, 3, 4, 5\}$ and $Y = \{2, 3, 2, 4, 5\}$ with correlation coefficient $r = 0.85$, standard deviations $s_x = 1.58$ and $s_y = 1.3$, and means $\bar{x} = 3$ and $\bar{y} = 3.2$, formulate the equation of the regression line.

Solution: First, calculate the slope: $m = r \times \frac{s_y}{s_x} = 0.85 \times \frac{1.3}{1.58} \approx 0.7$.

Next, calculate the y-intercept: $b = \bar{y} - m\bar{x} = 3.2 - 0.7 \times 3 = 1.1$.

So, the equation of the regression line is $y = 0.7x + 1.1$.

The following figure shows the graph of regression line.

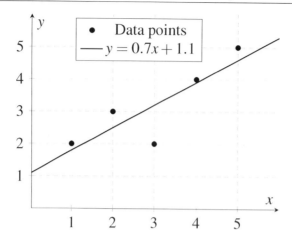

Example With a correlation coefficient of $r = -1$, standard deviations of $s_X = 1.58$, $s_Y = 1.58$, and means of $\overline{X} = 3$, $\overline{Y} = 3$ for the sets $X = \{1,2,3,4,5\}$ and $Y = \{5,4,3,2,1\}$, calculate the equation of the regression line.

Solution: We calculate the slope: $m = r \times \frac{s_Y}{s_X} = -1 \times \frac{1.58}{1.58} = -1$.

Next, we find the y-intercept: $b = \overline{Y} - m\overline{x} = 3 - (-1) \times 3 = 6$.

Thus, the regression line equation is $y = -x + 6$.

See the figure below for a graphical representation of points and regression line.

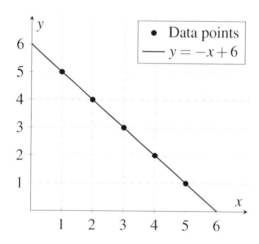

11.6 Understanding Correlation and Causation

The term 'Correlation' refers to a statistical relationship between two variables. A common misconception is that correlation implies a cause-effect relationship between these variables. This is not necessarily true and leads us to the concept of 'Causation'.

11.6 Understanding Correlation and Causation

'Correlation' implies a mutual relationship where changes in one variable coincide with changes in another. On the other hand, 'Causation' implies a cause-and-effect relationship where a change in one variable (the cause) produces a change in another variable (the effect).

Key Point

Correlation does not imply causation. A statistical relationship between variables, known as correlation, does not mean that one variable is the cause of changes in the other.

Even when correlation is high, it is possible that other, unseen factors are causing changes in both variables. Conversely, one variable could indeed cause changes in the other, but the correlation could be obscured by other factors.

Therefore, when interpreting a correlation, one should not automatically assume a cause-and-effect relationship.

However, in some cases, correlation can guide further investigation into a potential causation. Experimental studies, where researchers manipulate one variable and observe changes in another, provide the most valid means of establishing causation.

Example

Suppose a study shows a strong correlation between students' performance in school and the amount of breakfast consumed; students who eat a healthy breakfast tend to perform better in school. Does this imply that breakfast is the sole cause of their better performance?

Solution: No. While the data shows a correlation, we can not say that eating a healthy breakfast alone causes better performance in school. There could be other factors at play. For instance, students who have healthy breakfasts are also likely to have other beneficial lifestyle habits, like regular study routines and sufficient sleep. So it is not accurate to attribute the better school performance to breakfast alone without investigating further.

Example

A medical research study observes a correlation between walking barefoot and developing cold symptoms. Would it be correct to assert that walking barefoot causes colds?

Solution: No. While walking barefoot and cold symptoms are correlated, other factors may be causing both events. For example, people are more likely to walk barefoot and catch colds in cold winter weather. Therefore, it may be the cold weather (and not walking barefoot) which is causing the increase in cold symptoms. Correlation does not imply causation; other potential causes should be investigated.

11.7 Understanding Permutations and Combinations

Permutations and combinations are two concepts in mathematics that deal with counting and organizing items. While permutations focus on the arrangement of items where order is important, combinations are more concerned with the selection of items where order does not matter.

The formula for permutations $P(n,r)$ when selecting r items from n items is given by $\frac{n!}{(n-r)!}$ where $n!$ is the factorial of n, which is the product of all positive integers up to n.

Combinations, on the other hand, are denoted as $C(n,r)$ and the formula is $\frac{n!}{r!(n-r)!}$.

> **Key Point**
>
> In permutations, the order of selection is important whereas in combinations, order does not matter. The formula for permutations $P(n,r) = \frac{n!}{(n-r)!}$ and for combinations $C(n,r) = \frac{n!}{r!(n-r)!}$, where $n!$ denotes the factorial of n.

Example How many ways can a team of 3 be chosen from a group of 10 people?

Solution: Since the order does not matter, we use the formula for combinations. Here $n = 10$ and $r = 3$. Substituting these into the formula, we get: $C(10,3) = \frac{10!}{3!(10-3)!} = 120$.

This means there are 120 different ways to choose a team of 3 from a group of 10 people.

Example How many different ways can you arrange 4 books on a shelf?

Solution: The lineup of books matters here, so we use the permutation formula. Here $n = 4$ (the books) and $r = 4$ (we are arranging all the books). Thus: $P(4,4) = \frac{4!}{(4-4)!} = 24$.

Note that $0! = 1$. Hence, there are 24 different ways of arranging 4 books on the shelf.

11.8 Solving Probability Problems

Probability defines a measure for the likelihood of an event to happen. It is a number between 0 and 1. An event with a probability of 1 is a certainty, whereas an event with a probability of 0 is impossible.

11.9 Practices

> **Key Point**
>
> The probability of event A:
>
> $$P(A) = \frac{\text{number of desired outcomes}}{\text{total number of possible outcomes}}.$$

Independent Events: Two events A and B are considered independent if the occurrence of A does not affect the occurrence of B, and vice versa.

> **Key Point**
>
> When we multiply the probabilities of two independent events, we get the probability that both occur. This can be represented by $P(A \text{ and } B) = P(A) \times P(B)$, where A and B are independent events.

Example A bag contains 3 red balls and 2 green balls. What is the probability of drawing a red ball from the bag?

Solution: In this problem, the total number of outcomes is 5 (3 red balls and 2 green balls). The successful outcomes are represented by the red balls which are 3. Hence, $P(\text{red ball}) = \frac{3}{5}$.

Example A fair six-sided dice is rolled twice, what is the probability of rolling a 6 on the first roll and an odd number on the second roll?

Solution: The probability of rolling a 6 on a fair dice is $\frac{1}{6}$ and the probability of rolling an odd number (i.e., 1, 3, or 5) is $\frac{3}{6} = \frac{1}{2}$. Since these are independent events, we multiply those probabilities. Thus, $P(6 \text{ and odd}) = \frac{1}{6} \times \frac{1}{2} = \frac{1}{12}$.

11.9 Practices

True or False:

1) If the data set has an odd number of observations, the median is the average of the two middle numbers.

2) The mean of a data set is always an integer.

3) A data set may have more than one mode.

4) The range of a data set is always a positive number.

5) The mean, median, mode, and range are all measures of central tendency.

Fill in the Blank:

6) A pie graph represents 100% of the data, which corresponds to a full circle of _____ degrees.

7) In a pie graph, each sector represents a _____ of the total.

8) The sum of all the pie slices in a pie graph should always be _____.

9) The larger the angle or sector area in the pie graph, the _____ the data it represents.

10) To find the angle that a piece of data should occupy in the pie graph, it is calculated as the _____ times 360°.

True or False:

11) A scatter plot always reveals a positive correlation between two variables.

12) Each dot in a scatter plot corresponds to an observation.

13) Scatter plots are not useful to spot patterns, trends, and correlations in a set of data.

14) In a scatter plot, the x-axis and y-axis represent the two variables from the data set.

15) If the points on a scatter plot seem to have no pattern, this indicates a positive correlation.

Solve:

16) In a box containing 5 blue, 4 red and 3 green marbles, what is the probability of picking a red marble?

17) A fair coin is tossed twice, what is the probability of getting heads both times?

11.9 Practices

18) There are 4 black and 6 white socks in a drawer, what is the probability of randomly drawing a pair of white socks?

19) A bag contains 4 red, 3 blue, and 2 green marbles. What is the probability of drawing a blue marble, replacing it, then drawing a green one?

20) In a box of assorted donuts with 3 chocolate, 4 glaze, and 2 strawberry donuts, what is the probability of picking a strawberry donut first then a chocolate donut without replacement?

Solve:

21) A team consists of 4 team members. If there are 14 possible members, how many different teams can be formed?

22) How many different ways can 5 books be arranged on a shelf?

23) A committee of 3 members is to be chosen from a club of 9 members. How many different committees can be formed?

24) There are 7 different books. In how many ways can we arrange 4 of them on a shelf?

25) A menu has 10 different dishes. A customer wants to try out 4 dishes. In how many ways can the customer select the dishes?

True or False:

26) The value of a correlation coefficient is always in the range of -1.0 to 1.0.

27) A negative correlation coefficient means that there is no correlation between the variables.

28) The Pearson correlation coefficient is represented by r.

29) If two variables have a correlation coefficient of 0.95, that means they have a strong negative correlation.

30) A correlation coefficient quantifies the degree to which two variables are related.

Fill in the Blank:

31) In the regression equation $y = mx + b$, m represents the _____ and b represents the _____.

32) The formula for slope m of the regression line is _____.

33) In the formula for the y-intercept b, the term $\bar{x}$ refers to _____.

34) The equation of the regression line is _____, when $r = 1$, $s_x = s_y = 1$, $\bar{x} = \bar{y} = 0$.

35) The equation for calculating the y-intercept of the regression line is _____.

True or False:

36) If two variables are correlated, it means that one variable is the cause of changes in the other.

37) Experimental studies where researchers manipulate one variable and observe changes in another, do not provide a valid means of establishing causation.

38) If there is no correlation between two variables, it is certain that there is no causation between them.

39) Unseen factors can cause changes in both variables even when correlation is high.

40) If a cause-and-effect relationship exists, it is referred to as correlation.

Answer Keys

1) False

2) False

3) True

4) False

5) False

6) 360

7) proportion

8) 360°

9) larger

10) $\frac{\text{data value}}{\text{total data sum}}$

11) False

12) True

13) False

14) True

15) False

16) $P(red) = \frac{1}{3}$

17) $P(H \cap H) = \frac{1}{4}$

18) $P(W \cap W) = \frac{1}{3}$

19) $P(B \cap G) = \frac{2}{27}$

20) $P(S \cap C) = \frac{1}{12}$

21) 1001 teams

22) 120 ways

23) 84 committees

24) 840 ways

25) 210 ways

26) True

27) False

28) True

29) False

30) True

31) slope, y-intercept

32) $m = r\frac{s_y}{s_x}$

33) the mean of x

34) $y = x$

35) $b = \bar{y} - m\bar{x}$

36) False

37) False

38) False

39) True

40) False

Answers with Explanation

1) If the data set has an odd number of observations, the median is the middle number, not the average of the two middle numbers.

2) The mean of a data set is not always an integer. It can be a decimal or a fraction as it is the sum of all values divided by the count of values.

3) A data set can indeed have more than one mode. This occurs when more than one number appears the most number of times.

4) The range of a data set is not always positive. If the highest and lowest numbers in the set are the same, then the range is 0.

5) While the mean, median, and mode are measures of central tendency, or measures of the center of a data set, range is a measure of dispersion. It provides information about the spread of the data set.

6) A full circle corresponds to 360 degrees.

7) Each sector in a pie graph represents a proportion of the total data.

8) The sum of all sectors in a circle (pie graph) should always be 360 degrees.

9) The larger the angle or sector area in the pie graph, the larger the data it represents. This is how the proportions are visually depicted.

10) The angle that a piece of data should occupy in the pie graph is calculated as the ratio of the data value to the total data sum, multiplied by $360°$.

11) A scatter plot may reveal a positive, negative or no correlation. The type of correlation depends on the data and cannot be predetermined.

12) Each dot in a scatter plot corresponds to an individual observation in the data set.

11.9 Practices

13) Scatter plots are very useful in spotting patterns, trends, and correlations in a set of data. It gives a visual representation of the relationship between two variables.

14) In a scatter plot, the *x*-axis and *y*-axis represent the two variables from the data set, making it easier to spot patterns, trends, and correlations.

15) If there is no apparent pattern in the data points of a scatter plot, this usually indicates no correlation. A positive correlation has an upward trend.

16) Total marbles in the box $= 5 + 4 + 3 = 12$, but only 4 are red. Hence, probability is $\frac{4}{12} = \frac{1}{3}$.

17) Each coin toss is an independent event with 2 outcomes. The probability of getting heads both times is $\frac{1}{2} \times \frac{1}{2} = \frac{1}{4}$.

18) Firstly, we calculate all possible combinations by using the formula $C(10,2) = \frac{10!}{2!(10-2)!} = 45$. Now, for two white socks, the possible combinations are $C(6,2) = \frac{6!}{2!(6-2)!} = 15$. Hence: $P(W \cap W) = \frac{15}{45} = \frac{1}{3}$

19) Total marbles $= 4 + 3 + 2 = 9$, splitting into blue and green we get $\frac{3}{9} \times \frac{2}{9} = \frac{2}{27}$.

20) Picked strawberry from 9 donuts, then chocolate from remaining 8. Hence, $\frac{2}{9} \times \frac{3}{8} = \frac{1}{12}$.

21) We use the formula for combinations. Here $n = 14$ and $r = 4$. Substituting these into the formula, we get $C(14,4) = \frac{14!}{4!(14-4)!} = 1001$. So, there are 1001 different ways to form a team.

22) The lineup of books matters here, so we use the permutation formula. Here $n = 5$ (the books), and $r = 5$ (we are arranging all the books). Thus $P(5,5) = \frac{5!}{(5-5)!} = 120$. Hence, there are 120 different ways of arranging 5 books on the shelf.

23) We use the formula for combinations. Here $n = 9$ and $r = 3$. Substituting these into the formula, we get $C(9,3) = \frac{9!}{3!(9-3)!} = 84$. Thus, there are 84 different ways to form a committee.

24) The arrangement of books matters here, so we use the formula for permutations. The number of books to select from $n = 7$ and the number of books to arrange $r = 4$. Substituting these into the formula, we get $P(7,4) = \frac{7!}{(7-4)!} = 840$. Hence there are 840 ways to arrange 4 books out of 7 on the shelf.

25) Since the order of selecting dishes does not matter, we use the formula for combinations. Here $n = 10$

(dishes) and $r = 4$ (we are selecting 4 dishes). Thus $C(10,4) = \frac{10!}{4!(10-4)!} = 210$. Hence, there are 210 different ways of selecting 4 dishes out of 10.

26) Yes, the value of a correlation coefficient is always in the range of -1.0 to 1.0.

27) A negative correlation coefficient means that the variables move in opposite directions, not that there is no correlation.

28) Yes, the Pearson correlation coefficient is usually represented by the letter r.

29) If two variables have a correlation coefficient of 0.95, it means they have a strong positive correlation, not a negative one.

30) Yes, a correlation coefficient does quantify the degree to which two variables are related.

31) In this equation, m is the slope of the regression line and b is the y-intercept of the line.

32) Slope m is calculated as r times $\frac{s_y}{s_x}$, where r is the correlation coefficient, s_x is the standard deviation of x, and s_y is the standard deviation of y.

33) $\bar{x}$ refers to the mean (or average) of the x-values.

34) The slope $m = r\frac{s_y}{s_x} = 1$. The y-intercept $b = \bar{y} - m\bar{x} = 0$. So the equation of the line is $y = x$.

35) The y-intercept b is calculated by subtracting the product of the slope m and the mean of x-values from the mean of y-values.

36) Correlation does not necessarily imply causation. A correlation between two variables does not guarantee that one variable causes changes in the other.

37) Experimental studies do provide a solid basis for determining causation. By manipulating one variable and observing changes in another, we can often isolate cause-and-effect relationships.

38) Causation may still exist between two variables even in the absence of correlation. Other factors might obscure the correlation that would be evident if all other things were equal.

39) Yes, even when correlation is high, unseen factors could be causing changes in both variables. These

11.9 Practices

variables can be observed or unobserved alternates, and confounders among others.

40) A cause-and-effect relationship is referred to as 'causation'. 'Correlation' implies a mutual relationship where changes in one variable coincide with changes in another, but it does not necessarily indicate a cause-and-effect relationship.

12. Praxis Algebra I (5162) Test Review and Strategies

12.1 The Praxis Algebra I Test Review

The *Praxis Algebra I* examination is a specialized test tailored for individuals aiming to *teach algebra at the high school level* within the United States. This test is a crucial component of the *Praxis Series*, a suite of exams curated and conducted by the *Educational Testing Service (ETS)*. These exams play a significant role in the certification and licensure processes for educators across numerous states and professional licensing bodies.

Focused squarely on algebra, the Praxis Algebra I exam (5162) tests the candidate's grasp of fundamental algebraic principles, their ability to solve problems effectively, and their understanding of the instructional strategies required to teach algebra with clarity. The test encompasses approximately **60 multiple-choice questions** that must be answered within a **150 minutes** time slot, spanning various algebraic topics, such as:

- **Number and Quantity**
- **Functions**
- **Linear Equations and Inequalities**
- **Quadratic and Polynomial Equations**
- **Rational and Radical Equations**
- **Exponential and Logarithmic Equations**

- **Data Interpretation and Statistics**
- **Probability**

ETS periodically revises the Praxis exams to ensure they remain in step with evolving educational standards and teaching methodologies. As a result, the specific content and structure of the Algebra I test may be updated to reflect these changes. Test-takers are permitted to use specific models of *graphing calculators*. The Praxis Algebra I test has **no penalty for wrong answers**, encouraging candidates to attempt all questions without fear of a negative impact on their overall score.

To successfully pass the Praxis Algebra I exam, candidates must achieve a designated minimum score, which is determined by the state or jurisdiction where licensure is being sought. This passing score varies, underscoring the importance for candidates to verify the requirements specific to their area. Achieving this score signifies a candidate's readiness to teach algebra, meeting the established standards of knowledge and pedagogy.

12.2 Praxis Algebra I (5162) Test-Taking Strategies

Successfully navigating the Praxis Algebra I (5162) test requires not only a solid understanding of algebraic concepts but also effective problem-solving strategies. In this section, we explore a range of strategies to optimize your performance and outcomes on the Praxis Algebra I (5162) test. From comprehending the question and using informed guessing to finding ballpark answers and employing backsolving and numeric substitution, these strategies will empower you to tackle various types of math problems with confidence and efficiency.

 ### #1 Understand the Questions and Review Answers

Below are a set of effective strategies to optimize your performance and outcomes on the Praxis Algebra I (5162) test.

- **Comprehend the Question:** Begin by carefully reviewing the question to identify keywords and essential information.
- **Mathematical Translation:** Translate the identified keywords into mathematical operations that will enable you to solve the problem effectively.

- **Analyze Answer Choices:** Examine the answer choices provided and identify any distinctions or patterns among them.
- **Visual Aids:** If necessary, consider drawing diagrams or labeling figures to aid in problem-solving.
- **Pattern Recognition:** Look for recurring patterns or relationships within the problem that can guide your solution.
- **Select the Right Method:** Determine the most suitable strategies for answering the question, whether it involves straightforward mathematical calculations, numerical substitution (plugging in numbers), or testing the answer choices (backsolving); see below for a comprehensive explanation of these methods.
- **Verification:** Before finalizing your answer, double-check your work to ensure accuracy and completeness.

Let's review some of the important strategies in detail.

#2 Use Educated Guessing

This strategy is particularly useful for tackling problems that you have some understanding of but cannot solve through straightforward mathematics. In such situations, aim to eliminate as many answer choices as possible before making a selection. When faced with a problem that seems entirely unfamiliar, there's no need to spend excessive time attempting to eliminate answer choices. Instead, opt for a random choice before proceeding to the next question.

As you can see, employing direct solutions is the most effective approach. Carefully read the question, apply the math concepts you've learned, and align your answer with one of the available choices. Feeling stuck? Make your best-educated guess and move forward.

Never leave questions unanswered! Even if a problem appears insurmountable, make an effort to provide a response. If necessary, make an educated guess. Remember, you won't lose points for an incorrect answer, but you may earn points for a correct one!

#3 Ballpark Estimates

A *"ballpark estimate"* is a *rough approximation*. When dealing with complex calculations and numbers, it's easy to make errors. Sometimes, a small decimal shift can turn a correct answer into an incorrect one, no matter how many steps you've taken to arrive at it. This is where ballparking can be incredibly useful.

If you have an idea of what the correct answer might be, even if it's just a rough estimate, you can often eliminate a few answer choices. While answer choices typically account for common student errors and closely related values, you can still rule out choices that are significantly off the mark. When facing a multiple-choice question, deliberately look for answers that don't even come close to the ballpark. This strategy effectively helps eliminate incorrect choices during problem-solving.

 #4 Backsolving

A significant portion of questions on the Praxis Algebra I (5162) test are presented in multiple-choice format. Many test-takers find multiple-choice questions preferable since the correct answer is among the choices provided. Typically, you'll have four options to choose from, and your task is to determine the correct one. One effective approach for this is known as *"backsolving."*

As mentioned previously, solving questions directly is the most optimal method. Begin by thoroughly examining the problem, calculating a solution, and then matching the answer with one of the available choices. However, if you find yourself unable to calculate a solution, the next best approach involves employing *"backsolving."*

When employing backsolving, compare one of the answer choices to the problem at hand and determine which choice aligns most closely. Frequently, answer choices are arranCBEST in either ascending or descending order. In such cases, consider testing options B or C first. If neither is correct, you can proceed either up or down from there.

 #5 Plugging In Numbers

Using numeric substitution or *'plugging in numbers'* is a valuable strategy applicable to a wide array of math problems encountered on the Praxis Algebra I (5162) test. This approach is particularly helpful in simplifying complex questions, making them more manageable and comprehensible. By employing this strategy thoughtfully, you can arrive at the solution with ease.

The concept is relatively straightforward. Simply replace unknown variables in a problem with specific values. When selecting a number for substitution, consider the following guidelines:

- Opt for a basic number (though not overly basic). It's generally advisable to avoid choosing 1 (or even 0). A reasonable choice often includes selecting the number 2.

- Avoid picking a number already present in the problem statement.
- Ensure that the chosen numbers are distinct when substituting at least two of them.
- Frequently, the use of numeric substitution helps you eliminate some of the answer choices, so it's essential not to hastily select the first option that appears to be correct.
- When faced with multiple seemingly correct answers, you may need to opt for a different set of values and reevaluate the choices that haven't been ruled out yet.
- If your problem includes fractions, a valid solution might require consideration of either *the least common denominator (LCD)* or a multiple of the LCD.
- When tackling problems related to percentages, it's advisable to select the number 100 for numeric substitution.

It is Time to Test Yourself

It's time to refine your skills with a practice examination designed to simulate the Praxis Algebra I (5162) Test. Engaging with the practice tests will help you to familiarize yourself with the test format and timing, allowing for a more effective test day experience. After completing a test, use the provided answer key to score your work and identify areas for improvement.

Before You Start

To make the most of your practice test experience, please ensure you have:
- A pencil for marking answers on the answer sheet.
- A timer to manage pacing, replicating potential time constraints in other testing scenarios.

Please note the following important points as you prepare to take your practice test:
- It's okay to guess! There is no penalty for incorrect answers, so make sure to answer every question.
- After completing the test, review the answer key to understand any mistakes. This review is crucial for your learning and preparation.
- An answer sheet is provided for you to record your answers. Make sure to use it.
- For each multiple-choice question, you will be presented with possible choices. Your task is to choose the best one.

Good Luck! Your preparation and practice are the keys to success.

13. Practice Test 1

Praxis Algebra I Practice Test

Total number of questions: 60

Time: 150 Minutes

Calculator is permitted for Praxis Algebra I Test.

13.1 Practices

1) Which statement is incorrect for the function $f(x) = 4x^2 - 8x + 3$?

☐ A. The axis of symmetry of the function f is $x = 1$.

☐ B. The vertex of the function f is at $(1, -1)$.

☐ C. The zeros of f are 1 and $\frac{3}{4}$.

☐ D. The function opens upwards.

2) Which of the following is a factor of $8x^{10} - 14x^5 + 2x^4$?

☐ A. $2x^6 - 7x + 1$

13.1 Practices

☐ B. $x - 2$

☐ C. $2x^2 + 1$

☐ D. $4x^6 - 7x + 1$

3) Determine the equation of a horizontal line passing through the point $(4, -7)$.

☐ A. $y = -7$

☐ B. $x = -7$

☐ C. $y = 4$

☐ D. $x = 4$

4) A segment of an exponential function is depicted on a coordinate plane.

Which of the following correctly describes the domain of the segment shown?

☐ A. $y \geq 7$

☐ B. $-\infty < x < 3$

☐ C. $x \leq 2$

☐ D. $0 < y < 7.2$

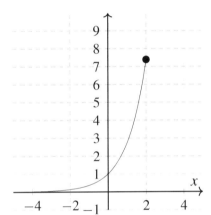

5) What is the product of $(6x + 3y)(4x + 3y)$?

☐ A. $24x^2 + 30xy + 9y^2$

☐ B. $3x^2 + 6xy + 3y^2$

☐ C. $6x^2 + 18xy + 3y^2$

☐ D. $10x^2 + 15xy + 9y$

6) Which of the following functions is equivalent to $g(x) = 6x^2 - 36x - 5$?

☐ A. $g(x) = 6(x - 3)^2 + 50$

☐ B. $g(x) = 6(x + 3)^2 - 50$

☐ C. $g(x) = 6(x - 3)^2 - 59$

☐ D. $g(x) = 6(x - 50)^2 - 3$

7) Among the given representations, which one shows y as a function of x?

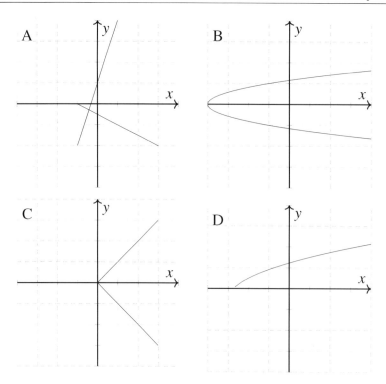

- ☐ A. Graph A
- ☐ B. Graph B
- ☐ C. Graph C
- ☐ D. Graph D

8) Which graph best represents the solution set of $y > \frac{1}{2}x - 2$?

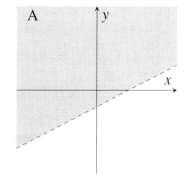
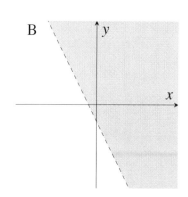

13.1 Practices

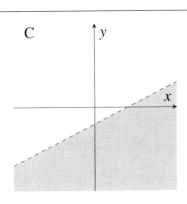
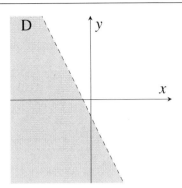

- ☐ A. Graph A
- ☐ B. Graph B
- ☐ C. Graph C
- ☐ D. Graph D

9) Given two characteristics of a quadratic function g:

   - The axis of symmetry of the graph of g is $x = -2$.
   - Function g has exactly one zero.

   Which graph could represent g?

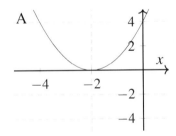
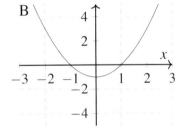
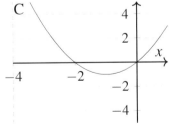
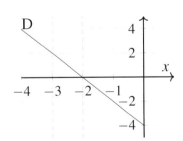

- ☐ A. Graph A
- ☐ B. Graph B
- ☐ C. Graph C
- ☐ D. Graph D

10) A freelancer works on two different projects. The total time spent on these projects cannot exceed 150 hours in a month.

Which option describes the solution set for all possible combinations of x, the number of hours spent on the first project, and y, the number of hours spent on the second project, in one month?

☐ A. A shaded region below and to the left of the line $x + y = 150$.

☐ B. A shaded region above and to the right of the line $x + y = 150$.

☐ C. A shaded region below and to the left of the line $x = 150$.

☐ D. A shaded region below and to the left of the line $y = 150$.

11) A blog's monthly views and comments are recorded in a table, indicating a linear relationship. Based on the data, what is the best prediction of the number of comments for each post if the number of monthly views reaches 6,000?

☐ A. 75

☐ B. 90

☐ C. 110

☐ D. 125

Monthly Views	Comments
3600	70
2400	50
4800	90
1200	30

12) A graph depicts the linear relationship between the volume of water in a tank and the time it has been filling.

13.1 Practices

Which of these best represents the rate of change of the water volume with respect to the time elapsed?

- A. $150 \frac{L}{hr}$
- B. $\frac{1}{150} \frac{L}{hr}$
- C. $250 \frac{L}{hr}$
- D. $\frac{1}{250} \frac{L}{hr}$

13) The graph of $h(x) = x^3$ is transformed to create the graph of $k(x) = 0.5h(x)$. Which graph best represents h and k?

Option A

Option B

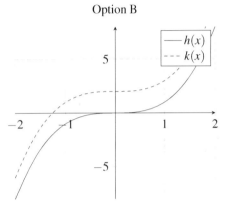

Option C

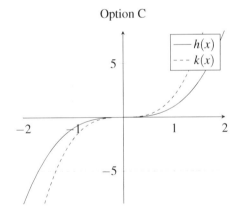

Option D

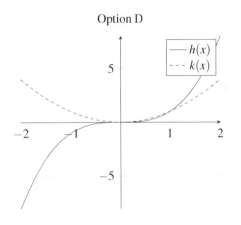

14) What is the value of the y-intercept of the graph $g(x) = 8.5(0.7)^x$? Write your answer in the box:

15) What is the ratio of the maximum value to the minimum value of the function $h(x) = -2x + 4$, where $-1 \leq x \leq 4$?

- A. $-\frac{5}{2}$
- B. $-\frac{3}{2}$
- C. $-\frac{10}{7}$

☐ D. $\frac{8}{3}$

16) A system of linear equations is represented by lines u and v. A table representing some points on line u and the graph of line v are provided.

Line u	
x	y
0	2
2	3
4	4

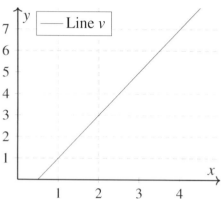

Which system of equations is best represented by lines u and v?

☐ A. $\begin{array}{l} 3x - 2y = 1 \\ x + y = 2 \end{array}$

☐ B. $\begin{array}{l} 3x - 2y = 1 \\ x - 2y = 4 \end{array}$

☐ C. $\begin{array}{l} y = \frac{2}{3}x - 2 \\ y = 2 - \frac{1}{3}x \end{array}$

☐ D. $\begin{array}{l} y = \frac{1}{2}x + 2 \\ y = 2x - 1 \end{array}$

17) In 2005, a programmer's income increased by $1,500 per year starting from a $35,000 annual salary. Which equation represents income greater than average?

(Let J represent income, y represent the number of years after 2005)

☐ A. $J > 1,500y + 35,000$

☐ B. $J > -1,500y + 35,000$

☐ C. $J < -1,500y + 35,000$

☐ D. $J < 1,500y - 35,000$

18) Solve: $\frac{4x+8}{x+4} \times \frac{x+4}{x+2} = $.

☐ A. 1

☐ B. 2

☐ C. 3

13.1 Practices

☐ D. 4

19) A graph shows the growth of a plant in height over several days.

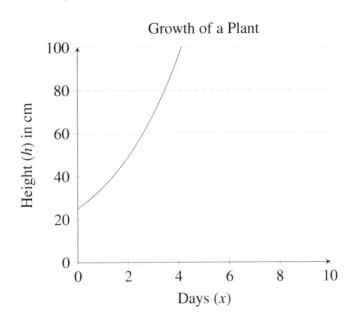

Based on this information, which function best describes the height of the plant h in centimeters per day?

☐ A. $h = 40(0.75)^x$

☐ B. $h = 25(1.4)^x$

☐ C. $h = 1.4(25)^x$

☐ D. $h = 0.75(40)^x$

20) Solve the following equation: $3x + 7 = 4x - 8$. Write the answer in the below box.

☐

21) Simplify the expression $\frac{8}{\sqrt{18}-4}$.

☐ A. $\sqrt{18} + 4$

☐ B. 3

☐ C. $4\sqrt{18} + 16$

☐ D. $4\sqrt{18}$

22) A golfer hits a ball, and the graph shows the height in meters of the golf ball above the ground as a quadratic function of d, the horizontal distance in meters from the golfer.

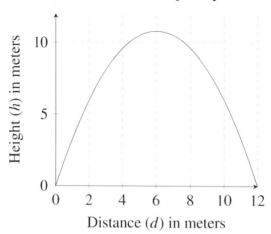

Golf Ball Trajectory

What is the domain of the function for this scenario?

☐ A. $0 \leq d \leq 12$

☐ B. $0 \leq h \leq 12$

☐ C. $2 \leq h \leq 3.5$

☐ D. $2 \leq d \leq 12$

23) Which of the following numbers is NOT a solution to the inequality $4x - 7 \leq 5x + 2$?

☐ A. -9

☐ B. -5

☐ C. -3

☐ D. -10

24) Which expression is equivalent to $16m^2 - 64$?

☐ A. $(4m-8)(4m-8)$

☐ B. $16(m-4)$

☐ C. $16m(m-4)$

☐ D. $16(m-2)(m+2)$

25) What is the value of the y-intercept of the graph of $h(x) = 15(1.3)^{x+2}$?

☐ A. 11.7

☐ B. 15

☐ C. 19.5

13.1 Practices

☐ D. 25.35

26) A nutritionist monitored the diet plan for a child for 24 weeks. The table and scatterplot show the calorie intake as a percentage of the child's daily requirement as a linear function of their age in weeks. What is the best prediction of the percentage of the daily calorie requirement that should be considered in the diet plan when the child is 30 weeks old?

Age (Weeks)	Calorie Intake (%)
0	100
4	102
8	105
12	107
16	109
20	112
24	115

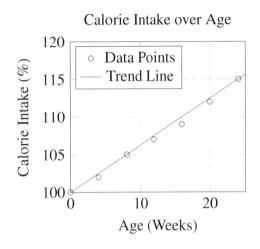

☐ A. 118%

☐ B. 120%

☐ C. 122%

☐ D. 125%

27) The graph of quadratic function h is displayed on a coordinate plane. What is the y-intercept of the graph of h? Write the answer in the below box.

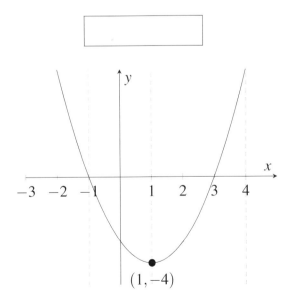

28) The line graphed on a coordinate plane represents the first of two equations in a system of linear equations. If the graph of the second equation in the system passes through the points $(0,-3)$ and $(4,5)$, which statement is true?

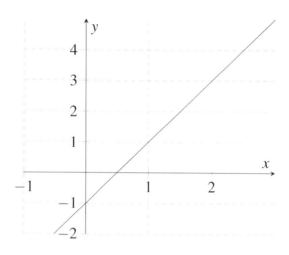

- ☐ A. The only solution to the system is $(-4,0)$.
- ☐ B. The only solution to the system is $(3,2)$.
- ☐ C. The system has no solution.
- ☐ D. The system has an infinite number of solutions.

29) Given $f(x) = x^2 - 7x + 12$, which statement is true about the zeroes of f?

- ☐ A. The zeroes are 4 and -3 because the factors of f are $(x-4)$ and $(x+3)$.
- ☐ B. The zeroes are -4 and 3 because the factors of f are $(x+4)$ and $(x-3)$.
- ☐ C. The zeroes are 4 and 3 because the factors of f are $(x-4)$ and $(x-3)$.
- ☐ D. The zeroes are -4 and -3 because the factors of f are $(x+4)$ and $(x+3)$.

30) A study shows the relationship between the amount of fertilizer used (in kilograms) and the yield of a crop (in tons). The data, represented by a quadratic function, is gathered over several seasons. Which function could best model this data?

13.1 Practices

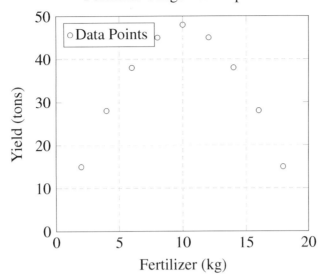

☐ A. $y = -x^2 + 20x$

☐ B. $y = x^2 - 20x$

☐ C. $y = -0.5x^2 + 10x$

☐ D. $y = 0.5x^2 - 10x$

31) Four tables display the relationship between temperature (in Celsius) and the solubility of a substance (in grams per 100ml of water). Which table does NOT represent solubility as a function of temperature?

☐ A.

Temperature (°C)	Solubility (g/100ml)
5	12
10	15
15	18
20	21

☐ B.

Temperature (°C)	Solubility (g/100ml)
5	14
10	16
15	18
20	20

	Temperature (°C)	Solubility (g/100ml)
	5	15
C.	5	16
	10	20
	15	24

	Temperature (°C)	Solubility (g/100ml)
	5	13
D.	10	17
	15	21
	20	25

32) If 40% of z is equal to 20% of 30, what is the value of $(z+5)^2$?

☐ A. 225

☐ B. 400

☐ C. 350

☐ D. 600

33) Consider a function $h(x)$ that has three distinct zeros. Which of the following could represent the graph of $h(x)$?

☐ A. A cubic graph crossing the x-axis at three different points.

☐ B. A quadratic graph touching the x-axis at one point and crossing at another.

☐ C. A linear graph crossing the x-axis at one point.

☐ D. A quartic graph touching the x-axis at two points and crossing at two others.

34) Determine the negative solution to the equation $2x^3 + 5x^2 - 12x = 0$.

35) A portion of a linear function h is graphed on a coordinate plane.

Which inequality best represents the domain of the part shown?

13.1 Practices

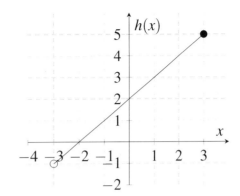

- [ ] A. $-3 < x \leq 3$
- [ ] B. $-3 \leq x < 3$
- [ ] C. $-1 < h(x) \leq 5$
- [ ] D. $-1 \leq h(x) < 5$

36) A linear function f has an x-intercept of 2 and a y-intercept of -3. Which option best describes f?

- [ ] A. A line crossing the x-axis at 2 and y-axis at -3.
- [ ] B. A line crossing the x-axis at -3 and y-axis at 2.
- [ ] C. A line with a positive slope crossing the x-axis at 2 and y-axis at -3.
- [ ] D. A line with a negative slope crossing the x-axis at 2 and y-axis at -3.

37) Given the system of equations: $4x+3y=7$ and $2x-6y=-4$, which of the following ordered pairs (x,y) satisfies both equations?

- [ ] A. $(2,1)$
- [ ] B. $(0,-1)$
- [ ] C. $(3,-2)$
- [ ] D. $(1,1)$

38) A line in the xy-plane passes through the point $(1,2)$ and has a slope of $\frac{1}{2}$. Which of the following points lies on this line?

- [ ] A. $(3,4)$
- [ ] B. $(4,1.5)$
- [ ] C. $(7,4.5)$
- [ ] D. $(5,4)$

39) What is the simplified form of $(5m^2+3m+4)-(3m^2-6)$?

- [ ] A. $2m^2+3m+10$
- [ ] B. $2m^2+3m-2$
- [ ] C. $2m+10$
- [ ] D. m^2+3m-2

40) Solve for x: $4(x+1) = 6(x-4) + 20$.

☐ A. 0

☐ B. 2

☐ C. 4

☐ D. 6.5

41) The given expression $2x - 3$ is a factor of which equation?

☐ A. $(2x - 3) + 17$

☐ B. $2x^3 - x^2 - 3x$

☐ C. $6(3 + 2x)$

☐ D. $2x^2 + x - 6$

42) What is the number of solutions to the equation $x^2 - 3x + 1 = x - 3$?

☐ A. 0

☐ B. 1

☐ C. 2

☐ D. Infinite

43) In the xy-plane, if $(0,0)$ is a solution to the system of inequalities $y < c - x$ and $y > x + b$, which of the following relationships between c and b must be true?

☐ A. $c < b$

☐ B. $c > b$

☐ C. $c = b$

☐ D. $c = b + c$

44) Calculate $f(5)$ for the following function f:

$$f(x) = x^2 - 3x.$$

☐ A. 5

☐ B. 10

☐ C. 15

☐ D. 20

13.1 Practices

45) John buys a pepper plant that is 5 inches tall. With regular watering, the plant grows 3 inches a year. Writing John's plant's height as a function of time, what does the y-intercept represent?

☐ A. The y-intercept represents the rate of growth of the plant which is 5 inches.

☐ B. The y-intercept represents the starting height of 5 inches.

☐ C. The y-intercept represents the rate of growth of the plant which is 3 inches per year.

☐ D. The y-intercept is always zero.

46) Multiply and write the product in scientific notation:

$$(3.1 \times 10^7) \times (1.8 \times 10^{-4}).$$

☐ A. 5.58×10^3

☐ B. 5.58×10^{11}

☐ C. 55.8×10^2

☐ D. 5.58×10^2

47) The first five terms in a geometric sequence are shown, where $x_1 = 3$.

3, −6, 12, −24, 48.

Based on this information, which equation can be used to find the nth term in the sequence, x_n?

☐ A. $x_n = 3(-2)^{n-1}$

☐ B. $x_n = 3^n$

☐ C. $x_n = -3n$

☐ D. $x_n = 3(-1)^n$

48) What is the parent graph of the following function and what transformations have taken place on it?
$y = 2x^2 - 8x + 6$

☐ A. The parent graph is $y = x^2$, which is stretched vertically by a factor of 2 and shifted 2 units right and 2 units down.

☐ B. The parent graph is $y = x^2$, which is shifted 4 units right and 6 units up.

☐ C. The parent graph is $y = x^2$, which is stretched vertically by a factor of 2 and shifted 4 units left.

☐ D. The parent graph is $y = x^2$, which is compressed vertically and shifted 2 units right and 6 units up.

49) A biology research group observes the growth of a certain algae species in a lake. The table shows the measured algae coverage area (in square meters) over several years since 2010. The data can be modeled by an exponential function.

Which function best models the data?

- A. $a(x) = 500(1.08)^x$
- B. $a(x) = 750(0.95)^x$
- C. $a(x) = 1.08(500)^x$
- D. $a(x) = 0.95(750)^x$

Year (since 2010)	Algae Coverage (sq. meters)
0	500
1	540
2	583.2
3	629.8
4	680.6

50) Determine the value of x in the solution of the system of equations:

$$4x + 5y = 17$$
$$2x - 3y = 3$$

- A. 1
- B. 2
- C. 3
- D. 4

51) Solve the following inequality: $|2x+4| \leq 6$.

- A. $x \leq 1$ or $x \geq -5$
- B. $-5 \leq x \leq 1$
- C. $x \leq -5$ or $x \geq 1$
- D. $-1 \leq x \leq 5$

52) How many ways can we pick a team of 4 people from a group of 10?

- A. 120
- B. 210
- C. 252
- D. 5040

53) The average of 7, 12, 25, and y is 15. What is the value of y?

13.1 Practices

- ☐ A. 11
- ☐ B. 20
- ☐ C. 15
- ☐ D. 16

54) In triangle DEF, if the measure of angle D is 45 degrees, what is the value of y? (Note: The figure is not drawn to scale)

- ☐ A. 40
- ☐ B. 50
- ☐ C. 55
- ☐ D. 58

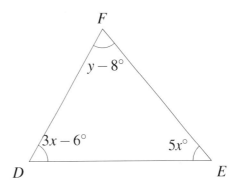

55) An angle is equal to one fifth of its complement. What is the measure of that angle?

- ☐ A. 18
- ☐ B. 10
- ☐ C. 15
- ☐ D. 36

56) When a number is subtracted from 35 and the difference is divided by that number, the result is 4. What is the value of the number?

- ☐ A. 3
- ☐ B. 5
- ☐ C. 7
- ☐ D. 9

57) Which of the following expressions is the inverse of the function $f(x) = \sqrt{x} + 2$?

- ☐ A. $(x+2)^2$
- ☐ B. $x^2 - 4x$
- ☐ C. $x^2 - 4x + 4$
- ☐ D. $x^2 - 4x + 2$

58) A chemical solution contains 8% alcohol. If there is 32 ml of alcohol, what is the volume of the solution?

- ☐ A. 300 ml
- ☐ B. 400 ml
- ☐ C. 500 ml
- ☐ D. 800 ml

59) What is the value of y in this equation?

$$5\sqrt{2y} + 7 = 27$$

- ☐ A. 4
- ☐ B. 20
- ☐ C. 50
- ☐ D. 8

60) A function $h(4) = 3$ and $h(6) = 5$. A function $k(5) = 7$ and $k(3) = 8$. What is the value of $k(h(6))$?

- ☐ A. 7
- ☐ B. 8
- ☐ C. 5
- ☐ D. 3

13.2 Answer Keys

1) C. The zeros of f are 1 and $\frac{3}{4}$.
2) D. $4x^6 - 7x + 1$
3) A. $y = -7$
4) C. $x \leq 2$
5) A. $24x^2 + 30xy + 9y^2$
6) C. $g(x) = 6(x-3)^2 - 59$
7) D. Graph D
8) A. Graph A
9) A.
10) A. A shaded region below and to the left of the line $x + y = 150$.
11) C. 110
12) A. $150\frac{L}{hr}$
13) A.
14) 8.5
15) B. $-\frac{3}{2}$
16) D. $y = \frac{1}{2}x + 2$
    $y = 2x - 1$
17) A. $J > 1,500y + 35,000$
18) D. 4
19) B. $h = 25(1.4)^x$
20) 15
21) C. $4\sqrt{18} + 16$
22) A. $0 \leq d \leq 12$
23) D. -10
24) D. $16(m-2)(m+2)$
25) D. 25.35
26) A. 118%
27) -3
28) C. The system has no solution.
29) C.
30) C. $y = -0.5x^2 + 10x$
31) C.
32) B. 400
33) A.
34) -4
35) A. $-3 < x \leq 3$
36) C. A line with a positive slope crossing the x-axis at 2 and y-axis at -3.
37) D. $(1,1)$
38) D. $(5,4)$
39) A. $2m^2 + 3m + 10$
40) C. 4
41) D. $2x^2 + x - 6$
42) B. 1
43) B. $c > b$
44) B. 10
45) B.
46) A. 5.58×10^3
47) A. $x_n = 3(-2)^{n-1}$
48) A.
49) A. $a(x) = 500(1.08)^x$
50) C. 3
51) B. $-5 \leq x \leq 1$
52) B. 210
53) D. 16

54) D. 58

55) C. 15

56) C. 7

57) C. $x^2 - 4x + 4$

58) B. 400 ml

59) D. 8

60) A. 7

13.3 Answers with Explanation

1) The function $f(x) = 4x^2 - 8x + 3$ can be rewritten as $f(x) = 4(x-1)^2 - 1$.

$$\text{Axis of symmetry} = 1 \Rightarrow \text{A is correct.}$$

$$\text{Vertex} = (1, -1) \Rightarrow \text{B is correct.}$$

$$a = 4 > 0 \quad \Rightarrow \text{Opens upwards} \Rightarrow \text{D is correct.}$$

The zeros of the function are found by solving $4x^2 - 8x + 3 = 0$. By the quadratic formula, the zeros (which are $\frac{1}{2}$ and $\frac{3}{2}$), do not match the values in option C.

Thus, option C is incorrect.

2) The polynomial $8x^{10} - 14x^5 + 2x^4$ can be factored as:

$$2x^4(4x^6 - 7x + 1).$$

Therefore, D is the correct choice.

3) A horizontal line has the form $y = k$, where k is a constant. The given point is $(4, -7)$, so the line must have a constant y-value of -7. Thus, the equation of the line is $y = -7$, which corresponds to option A.

4) The domain of a function refers to the set of all possible x values for which the function is defined. In the graph, the exponential function segment is shown continuing leftwards to $-\infty$ and ending at $x = 2$, where it is marked by a filled circle. This indicates that the function is defined for all x values up to and including 2. Therefore, the correct description of the domain of the segment shown is $x \leq 2$, as it includes all x values less than or equal to 2, making option C correct.

5) Multiplying the polynomials $(6x + 3y)$ and $(4x + 3y)$ gives:

$$(6x + 3y)(4x + 3y) = 24x^2 + 18xy + 12xy + 9y^2 = 24x^2 + 30xy + 9y^2.$$

Therefore, option A is the correct answer.

6) To find the equivalent form of $g(x) = 6x^2 - 36x - 5$, complete the square:

$$g(x) = 6(x^2 - 6x) - 5 = 6\left[(x-3)^2 - 3^2\right] - 5 = 6(x-3)^2 - 6 \times 9 - 5.$$

Simplifying gives:

$$g(x) = 6(x-3)^2 - 54 - 5 = 6(x-3)^2 - 59.$$

Therefore, option C is the correct equivalent function.

7) To determine which graph represents y as a function of x, we need to check if each x value corresponds to exactly one y value. This is commonly known as the "vertical line test," where a vertical line drawn at any x-coordinate should intersect the graph at no more than one point.

- Graph A shows two linear functions intersecting, meaning at some x values, there are two corresponding y values. It fails the vertical line test.

- Graph B depicts a sideways parabola. This graph fails the vertical line test because a vertical line can intersect the graph at two points in certain areas.

- Graph C shows two intersecting linear functions, similar to Graph A, and also fails the vertical line test.

- Graph D represents a square root function. Every x value in its domain corresponds to exactly one y value, successfully passing the vertical line test.

Therefore, Graph D is the only one that correctly represents y as a function of x, making option D correct.

8) To determine which graph represents the solution set of the inequality $y > \frac{1}{2}x - 2$, we need to identify the graph that correctly depicts this relationship.

- Graph A shows the line $y = \frac{1}{2}x - 2$ as a dashed line, indicating that the line itself is not included in the solution set. The shaded area above the line represents all the (x, y) points where y is greater than $\frac{1}{2}x - 2$. This correctly represents the solution set of the given inequality.

- Graph B depicts a line with a negative slope and a different inequality, which is not relevant to the given inequality $y > \frac{1}{2}x - 2$.

- Graph C, similar to Graph A, shows the line $y = \frac{1}{2}x - 2$, but the shading is below the line, representing $y < \frac{1}{2}x - 2$, which is not the inequality we are looking to represent.

- Graph D, like Graph B, represents a different inequality not related to $y > \frac{1}{2}x - 2$.

Therefore, Graph A correctly represents the solution set for $y > \frac{1}{2}x - 2$, making option A correct.

13.3 Answers with Explanation

9) To determine which graph could represent the quadratic function g with the given characteristics, we need to analyze the provided graphs based on two criteria: the axis of symmetry and the number of zeros.

- The axis of symmetry of a quadratic function is a vertical line that divides the graph into two mirror-image halves. For function g, the axis of symmetry is given as $x = -2$.

Other options have not both of these properties.

Therefore, Graph A correctly matches both the axis of symmetry at $x = -2$ and the presence of exactly one zero, making option A the correct choice.

10) The inequality representing the condition is $x + y \leq 150$. This inequality means the combined hours x and y should not exceed 150. Therefore, the solution set is represented by the region below and to the left of the line $x + y = 150$, which is option A.

11) Despite the rearrangement, the relationship between monthly views and comments is still linear. To find the number of comments for 6,000 views, we first calculate the slope (m) of the linear relationship.

Using two points from the table, for example, (3600, 70) and (1200, 30), we find the slope:

$$m = \frac{70 - 30}{3600 - 1200} = \frac{40}{2400} = \frac{1}{60}.$$

The linear equation is $y = mx + b$. Using the point (1200, 30) to solve for 'b':

$$30 = \frac{1}{60} \times 1200 + b \Rightarrow b = 30 - 20 = 10.$$

Thus, the equation is $y = \frac{1}{60}x + 10$. For 6,000 views:

$$y = \frac{1}{60} \times 6000 + 10 = 100 + 10 = 110.$$

Therefore, the correct answer is option C.

12) The rate of change of the water volume with respect to time can be determined by calculating the slope of the line on the graph. The slope represents the change in volume per unit time, which is the rate at which the tank is being filled.

From the graph, we can take two points to calculate the slope. For instance, using the points (1 hr, 150 L) and (2 hrs, 300 L):

$$\text{Slope} = \frac{\text{Change in Volume}}{\text{Change in Time}} = \frac{300L - 150L}{2\,hr - 1\,hr} = \frac{150L}{1\,hr} = 150\frac{L}{hr}.$$

Therefore, the rate of change of the water volume with respect to the time elapsed is $150\frac{L}{hr}$, making option A correct.

13) The transformation from $h(x) = x^3$ to $k(x) = 0.5h(x)$ involves scaling the graph of $h(x)$ vertically by a factor of 0.5. This means that the heights of the points on $h(x)$ are halved to get the corresponding points on $k(x)$, resulting in a vertically scaled (compressed) version of the cubic function. Therefore, option A is correct.

14) The y-intercept of a function is the value of y when $x = 0$. For the function $g(x) = 8.5(0.7)^x$, substituting $x = 0$ gives:

$$g(0) = 8.5(0.7)^0 = 8.5 \times 1 = 8.5.$$

Therefore, the y-intercept of the graph is 8.5.

15) For the function $h(x) = -2x + 4$, evaluate at the endpoints of the interval:

$$h(-1) = -2(-1) + 4 = 2 + 4 = 6 \quad \text{(Maximum value)},$$

$$h(4) = -2(4) + 4 = -8 + 4 = -4 \quad \text{(Minimum value)}.$$

The ratio of the maximum value to the minimum value is:

$$\frac{\text{Maximum}}{\text{Minimum}} = \frac{6}{-4} = -\frac{3}{2}.$$

Therefore, the correct answer in option B.

16) To determine which system of equations is best represented by lines u and v, we need to derive the equations for these lines based on the given information.

- For line u, we can use the points from the table to find its equation. With points $(0, 2)$ and $(2, 3)$, the slope m is:

$$m = \frac{3-2}{2-0} = \frac{1}{2}.$$

13.3 Answers with Explanation

Using the point-slope form $y - y_1 = m(x - x_1)$ with point $(0, 2)$:

$$y - 2 = \frac{1}{2}(x - 0) \Rightarrow y = \frac{1}{2}x + 2.$$

- For line v, the equation is provided in the graph's legend: $y = 2x - 1$.

Comparing these derived equations with the provided options, we find that option D, which states:

$$y = \tfrac{1}{2}x + 2$$
$$y = 2x - 1$$

perfectly matches the equations for lines u and v, making it the correct choice.

17) The equation for the programmer's income as a function of years after 2005 is $J = 1,500y + 35,000$, where $1,500$ is the annual increase and $35,000$ is the starting salary. To represent income greater than the average, the inequality would be $J > 1,500y + 35,000$, making option A correct.

18) Simplify the expression:
$$\frac{4x + 8}{x + 4} \times \frac{x + 4}{x + 2} = \frac{4(x + 2)}{x + 4} \times \frac{x + 4}{x + 2}.$$

The factors $x + 4$ and $x + 2$ cancel out, leaving: 4, option D.

19) The graph of the function $h = 25(1.4)^x$ shows a pattern of exponential growth, which is typical for the height of a plant over time. In this function, 25 represents the initial height of the plant, and 1.4 is the growth factor per day. As x (days) increases, the height h increases at an exponential rate, indicating that the plant grows faster as time progresses. This type of growth is common in many biological processes, making Option B the best representation of the plant's growth. The other options either depict a decreasing growth pattern or an unrealistic exponential increase.

20) To solve the equation $3x + 7 = 4x - 8$, we first move all x terms to one side and the constants to the other side:

$$3x - 4x = -8 - 7$$

$$-x = -15 \Rightarrow x = \frac{-15}{-1} = 15.$$

Therefore, the solution of the equation is $x = 15$.

21) To simplify $\frac{8}{\sqrt{18}-4}$, use the conjugate to rationalize the denominator:

$$\frac{8}{\sqrt{18}-4} \times \frac{\sqrt{18}+4}{\sqrt{18}+4} = \frac{8(\sqrt{18}+4)}{18-16}.$$

Simplifying further:

$$\frac{8\sqrt{18}+32}{2} = 4\sqrt{18}+16.$$

Therefore, the correct answer is option C.

22) The graph of the quadratic function representing the golf ball's trajectory shows the height of the ball as it travels horizontally. The domain of this function, d, represents the horizontal distance the ball covers from the golfer. In this scenario, the domain is given as $0 \leq d \leq 12$ meters, which is Option A. The other options (B, C, D) refer to the range of heights or incorrect distance ranges, which are not appropriate to describe the domain of the function in this context.

23) Solving the inequality $4x - 7 \leq 5x + 2$:

$$4x - 7 \leq 5x + 2 \Rightarrow -7 - 2 \leq 5x - 4x \Rightarrow -9 \leq x.$$

Therefore, x must be greater than or equal to -9. The number -10 does not satisfy this condition, making option D the correct choice.

24) The expression $16m^2 - 64$ is a difference of squares, which can be factored as:

$$16m^2 - 64 = 16(m^2 - 4) = 16(m-2)(m+2).$$

Therefore, option D is the correct equivalent expression.

25) The y-intercept of a function is found by setting $x = 0$. For the function $h(x) = 15(1.3)^{x+2}$:

$$h(0) = 15(1.3)^{0+2} = 15 \times 1.3^2 = 15 \times 1.69 = 25.35.$$

Therefore, the correct answer is option D.

13.3 Answers with Explanation

26) The scatter plot shows a linear trend in the child's calorie intake as a percentage of their daily requirement over time. The trend line approximates this relationship, and we can use it to predict the calorie intake at 30 weeks. Based on the trend line's equation derived from the scatter plot, the best prediction for 30 weeks can be calculated. If the trend line equation is approximately $y = 100 + 0.625x$, then at $x = 30$ weeks, the prediction is $y = 100 + 0.625 \times 30 = 118.75\%$. This value is closest to Option A, 118%.

27) To determine the y-intercept of the graph of the quadratic function h, which we do not have in explicit form, we can utilize the given points on the graph: $(1, -4)$, $(-1, 0)$, and $(3, 0)$. The y-intercept occurs where the graph crosses the y-axis, which is at $x = 0$.

Since we know that the points $(-1, 0)$ and $(3, 0)$ are zeros of the function, the quadratic function h can be expressed in its factored form:
$$h(x) = a(x+1)(x-3),$$
where a is a constant.

To find a, we use the third point $(1, -4)$:
$$-4 = a(1+1)(1-3) = a \times 2 \times (-2) = -4a.$$

Solving for a gives us $a = 1$.

Now we have the function:
$$h(x) = (x+1)(x-3).$$

To find the y-intercept, substitute $x = 0$ into $h(x)$:
$$h(0) = (0+1)(0-3) = 1 \times (-3) = -3.$$

Therefore, the y-intercept of the graph is -3.

28) The first line's equation, as represented in the graph, appears to be $y = 2x - 1$.

The second line passes through the points $(0, -3)$ and $(4, 5)$. We can find its slope using these points:
$$\text{Slope} = \frac{5 - (-3)}{4 - 0} = \frac{8}{4} = 2.$$

Since one of the points is $(0, -3)$, which is the y-intercept, the equation of the second line can be written as:

$$y = 2x - 3.$$

Comparing the two lines: - The first line is $y = 2x - 1$. - The second line is $y = 2x - 3$.

Both lines have the same slope but different y-intercepts. Therefore, they are parallel and do not intersect. When two lines in a system of linear equations are parallel and have different y-intercepts, they do not have any points of intersection, meaning there is no solution to the system. Hence, the correct answer is option C, indicating that the system has no solution.

29) To determine the zeroes of the quadratic function $f(x) = x^2 - 7x + 12$, we can factorize the quadratic equation. The zeroes are the values of x that make $f(x) = 0$.

The factorization of $f(x)$ involves finding two numbers that multiply to 12 (the constant term) and add up to -7 (the coefficient of the x term). These numbers are -4 and -3, as $-4 \times -3 = 12$ and $-4 + -3 = -7$. Therefore, the equation can be factored as:

$$f(x) = x^2 - 7x + 12 = (x-4)(x-3).$$

Setting each factor equal to zero gives the zeroes of the function:

$$x - 4 = 0 \Rightarrow x = 4,$$

$$x - 3 = 0 \Rightarrow x = 3.$$

Thus, the zeroes of the function are 4 and 3. These correspond to the factors $(x-4)$ and $(x-3)$. Therefore, the correct option is C: "The zeroes are 4 and 3 because the factors of f are $(x-4)$ and $(x-3)$."

30) The scatter plot shows a quadratic trend where the yield initially increases with the amount of fertilizer and then decreases after reaching a peak. Among the given options, the function $y = -0.5x^2 + 10x$ best models this trend. It has a positive linear term, indicating an initial increase in yield with increasing fertilizer, and a negative quadratic term, reflecting the decrease in yield beyond a certain point. Note that the option A is incorrect because for 10 kg of fertilizer, we have approximately 50 tons of the crop.

13.3 Answers with Explanation

31) To determine which table does not represent solubility as a function of temperature, we need to identify a table where a single temperature value corresponds to more than one solubility value.

- Tables A, B, and D all have unique solubility values for each temperature, which is consistent with the definition of a function.

- Table C shows the temperature of 5°C corresponding to two different solubility values (15 and 16 g/100ml). This violates the definition of a function, where each input (temperature) should map to exactly one output (solubility).

Therefore, Table C is the one that does NOT represent solubility as a function of temperature.

32) Solve the equation $0.40z = 0.20 \times 30$:

$$0.40z = 6 \Rightarrow z = \frac{6}{0.40} = 15.$$

Then calculate $(z+5)^2$ with $z = 15$:

$$(15+5)^2 = 20^2 = 400.$$

Thus, the value of $(z+5)^2$ is 400.

33) A function with three distinct zeros will cross the x-axis at three separate points. This is typically characteristic of a cubic function, as cubic functions can have up to three real zeros. Therefore, a cubic graph that crosses the x-axis at three different points would represent $h(x)$.

34) To find the negative solution to the equation $2x^3 + 5x^2 - 12x = 0$, we first factor out the common term, which in this case is x:

$$x(2x^2 + 5x - 12) = 0.$$

The equation now has a factored term $x(2x^2 + 5x - 12)$. The roots of the equation are the solutions to $x = 0$ and $2x^2 + 5x - 12 = 0$. The solution $x = 0$ is already evident. To find the other solutions, we focus on the quadratic equation $2x^2 + 5x - 12 = 0$.

We can attempt to factorize the quadratic or use the quadratic formula. The quadratic formula is given by:

$$x = \frac{-b \pm \sqrt{b^2 - 4ac}}{2a},$$

where $a = 2$, $b = 5$, and $c = -12$. Applying these values:

$$x = \frac{-5 \pm \sqrt{5^2 - 4 \times 2 \times -12}}{2 \times 2} = \frac{-5 \pm \sqrt{25 + 96}}{4} = \frac{-5 \pm \sqrt{121}}{4}.$$

This simplifies to:
$$x = \frac{-5 \pm 11}{4}.$$

The two solutions from this are:
$$x_1 = \frac{-5 + 11}{4} = \frac{6}{4} = 1.5,$$
$$x_2 = \frac{-5 - 11}{4} = \frac{-16}{4} = -4.$$

Among these, the negative solution is $x = -4$, which is the final answer.

35) The domain of a function represents the set of all possible input values (in this case, x values) for which the function is defined. In the graph, the portion of the linear function h shown runs from just beyond $x = -3$ to and including $x = 3$. This is represented by the inequality $-3 < x \leq 3$, where x is greater than -3 but less than or equal to 3. Therefore, the correct domain for the part of the function shown is Option A: $-3 < x \leq 3$.

36) The linear function f with an x-intercept of 2 and a y-intercept of -3 is best represented by a line that crosses the x-axis at 2 (point $(2,0)$) and the y-axis at -3 (point $(0,-3)$). Since the line moves from lower left to upper right, it has a positive slope. This matches Option C: A line with a positive slope crossing the x-axis at 2 and the y-axis at -3.

37) Solving the system by substitution or elimination method:

$$4x + 3y = 7$$
$$2x - 6y = -4$$

Multiply the second equation by -2 and add to the first equation:

$$4x + 3y = 7$$
$$-4x + 12y = 8$$

13.3 Answers with Explanation

Adding these gives:
$$15y = 15 \Rightarrow y = 1.$$

Substitute y into one of the equations:
$$4x + 3(1) = 7 \Rightarrow x = 1.$$

Therefore, the ordered pair that satisfies both equations is $(1,1)$, which is option D.

38) Using the slope-point form of a line:
$$y - y_1 = m(x - x_1)$$

Substitute $(1,2)$ and slope $\frac{1}{2}$:
$$y - 2 = \frac{1}{2}(x - 1)$$

Simplify to get the equation of the line:
$$y = \frac{1}{2}x + \frac{3}{2}$$

Check each point to see which lies on the line. For $(5,4)$:
$$4 = \frac{1}{2}(5) + \frac{3}{2} = \frac{5}{2} + \frac{3}{2} = 4$$

Thus, point $(5,4)$ lies on the line, making option D correct.

39) Simplify the expression by subtracting the second polynomial from the first:
$$(5m^2 + 3m + 4) - (3m^2 - 6) = 5m^2 + 3m + 4 - 3m^2 + 6 = 2m^2 + 3m + 10.$$

Therefore, the simplified form is $2m^2 + 3m + 10$, which is option A.

40) Expand and simplify the equation:
$$4x + 4 = 6x - 24 + 20,$$

$$4x+4 = 6x-4.$$

Bring all x terms to one side and constants to the other side:

$$4x - 6x = -4 - 4 \Rightarrow -2x = -8 \Rightarrow x = \frac{-8}{-2} = 4.$$

The solution is $x = 4$, making option C correct.

41) To determine if $2x - 3$ is a factor, it must divide one of the given equations without a remainder. We have:

$$2x^2 + x - 6 = (2x - 3)(x + 2).$$

Therefore, the correct answer in option D.

42) Rearrange the equation to form a quadratic equation:

$$x^2 - 3x + 1 = x - 3 \Rightarrow x^2 - 4x + 4 = 0.$$

Factorize or use the quadratic formula:

$$(x-2)^2 = 0$$

The solution is $x = 2$ indicating only one unique solution, making option C correct.

43) Substituting $(0,0)$ into the inequalities:

$$0 < c - 0 \Rightarrow c > 0,$$

$$0 > 0 + b \Rightarrow b < 0.$$

Therefore, c must be greater than b for both inequalities to hold true at $(0,0)$, making option B correct.

44) Substitute $x = 5$ into the function:

$$f(5) = 5^2 - 3 \times 5 = 25 - 15 = 10.$$

13.3 Answers with Explanation

Therefore, $f(5)$ equals 10, making option B correct.

45) In a linear function representing growth over time, the y-intercept represents the initial value of the variable being measured when time is zero. In this case, when time is zero (at the start), the height of the plant is 5 inches. Therefore, the y-intercept represents the starting height of the plant, which is 5 inches, making option B correct.

46) Multiply the numbers and add the exponents:

$$(3.1 \times 10^7) \times (1.8 \times 10^{-4}) = 5.58 \times 10^{7-4} = 5.58 \times 10^3.$$

Therefore, the product in scientific notation is 5.58×10^3, which is option A.

47) The sequence shows a pattern of each term being multiplied by -2 to get the next term. The first term is 3, and each subsequent term is obtained by multiplying the previous term by -2. The nth term of a geometric sequence is given by $x_1 \times r^{(n-1)}$, where x_1 is the first term and r is the common ratio. Thus, the formula for the nth term is:

$$x_n = 3(-2)^{n-1}.$$

Therefore, option A is correct.

48) The parent graph is $y = x^2$. The given function $y = 2x^2 - 8x + 6$ can be rewritten in vertex form as $y = 2(x-2)^2 - 2$. This indicates that the graph of $y = x^2$ has been transformed as follows:

1. Stretched vertically by a factor of 2 (due to the coefficient 2 in front of $(x-2)^2$).
2. Shifted 2 units to the right (due to the $(x-2)$ term).
3. Shifted 2 units down (due to the -2 outside the square).

Therefore, option A is correct.

49) The data in the table shows algae coverage increasing each year, which suggests exponential growth. In an exponential growth model, the value increases by a constant percentage each year. Option A ($a(x) = 500(1.08)^x$) represents an exponential function where the initial coverage is 500 sq. meters, and it grows by 8% each year (as indicated by the factor 1.08). This model aligns with the pattern observed in the table, where each year's coverage is approximately 8% more than the previous year.

The other options either represent a decrease (due to factors less than 1) or an unrealistic exponential increase (with the base as 500 or 750). Therefore, Option A is the best model for the data.

50) To solve the system
$$4x + 5y = 17$$
$$2x - 3y = 3,$$
we can use the elimination method. Multiply the second equation by 2 to align the coefficients of x:
$$4x + 5y = 17$$
$$4x - 6y = 6.$$

Subtract the second equation from the first:
$$11y = 11.$$

This gives $y = 1$. Substitute $y = 1$ into the first equation:
$$4x + 5 = 17 \Rightarrow 4x = 12 \Rightarrow x = 3.$$

Thus, $x = 3$, option C.

51) To solve the inequality $|2x+4| \leq 6$, we consider two cases based on the absolute value:

1. $2x + 4 \leq 6$
$$2x \leq 2 \Rightarrow x \leq 1.$$

2. $-(2x+4) \leq 6$ which simplifies to $2x + 4 \geq -6$
$$2x \geq -10 \Rightarrow x \geq -5.$$

Combining both conditions, we get the solution set:
$$-5 \leq x \leq 1.$$

This is represented by option B.

13.3 Answers with Explanation

52) To find the number of ways to pick a team of 4 people from a group of 10, we use the combination formula, which is given by:

$$\binom{n}{r} = \frac{n!}{r!(n-r)!},$$

where n is the total number of people, r is the number of people to choose, and ! denotes the factorial.

For our case, $n = 10$ and $r = 4$, so the formula becomes:

$$\binom{10}{4} = \frac{10!}{4!(10-4)!} = \frac{10 \times 9 \times 8 \times 7}{4 \times 3 \times 2 \times 1} = 210.$$

Therefore, there are 210 ways to pick a team of 4 people from a group of 10, which is option B.

53) The average of the numbers is calculated by dividing the sum of the numbers by the count of the numbers. Given that the average of the numbers 7, 12, 25, and y is 15, we can set up the equation:

$$\frac{7+12+25+y}{4} = 15.$$

Simplifying the equation:

$$44 + y = 15 \times 4,$$

$$44 + y = 60,$$

$$y = 60 - 44 = 16.$$

Therefore, the value of y is 16, which is option D.

54) Given that angle D is $45°$ and is represented as $3x - 6°$, we first find the value of x:

$$3x - 6 = 45.$$

Solving for x:

$$3x = 45 + 6 \Rightarrow x = 17.$$

Now, using the value of x to find y, from the angle sum property of a triangle, which states that the sum of angles in a triangle is $180°$:

$$45° + (5 \times 17)° + (y - 8)° = 180°.$$

Simplify and solve for y:

$$45 + 85 + y - 8 = 180 \Rightarrow y = 58.$$

Therefore, the value of y is 58°.

55) Let the angle be x degrees. The complement of an angle is 90 degrees minus the angle. Since the angle is one fifth of its complement, we can set up the equation:

$$x = \frac{1}{5}(90 - x).$$

Solving for x:

$$x = 18 - \frac{1}{5}x \Rightarrow \frac{6}{5}x = 18 \Rightarrow x = \frac{18 \times 5}{6} = 15.$$

Therefore, the measure of the angle is 15 degrees, making option C correct.

56) Let the number be x. According to the problem, when x is subtracted from 35 and the difference is divided by x, the result is 4. This can be written as an equation:

$$\frac{35 - x}{x} = 4.$$

Solving for x:

$$35 - x = 4x \Rightarrow 35 = 5x \Rightarrow x = \frac{35}{5} = 7.$$

Therefore, the value of the number is 7, which is option C.

57) To find the inverse of the function $f(x) = \sqrt{x} + 2$, we first replace $f(x)$ with y:

$$y = \sqrt{x} + 2.$$

To find the inverse, we interchange x and y and solve for y:

$$x = \sqrt{y} + 2.$$

Isolating $\sqrt{y}$:

$$\sqrt{y} = x - 2.$$

13.3 Answers with Explanation

Squaring both sides to remove the square root:

$$y = (x-2)^2 = x^2 - 4x + 4.$$

Therefore, the inverse function is $y = x^2 - 4x + 4$, which matches option C.

58) To find the total volume of the solution, we can set up a proportion based on the percentage of alcohol. Since the solution is 8% alcohol, and there are 32 ml of alcohol, we can write the equation:

$$\frac{8}{100} = \frac{32}{\text{Total Volume}}.$$

Solving for the total volume:

$$\text{Total Volume} = \frac{32}{0.08} = 400.$$

Therefore, the total volume of the solution is 400 ml, which is option B.

59) To solve the equation $5\sqrt{2y} + 7 = 27$, first isolate the radical expression:

$$5\sqrt{2y} = 27 - 7 = 20.$$

Divide both sides by 5:

$$\sqrt{2y} = \frac{20}{5} = 4.$$

Square both sides to remove the square root:

$$2y = 4^2 = 16.$$

Finally, solve for y:

$$y = \frac{16}{2} = 8.$$

Therefore, the value of y is 8.

60) To find $k(h(6))$, we first evaluate $h(6)$ and then apply the result to the function k.

First, evaluate $h(6)$:

$$h(6) = 5.$$

Next, use this result in k:

$$k(h(6)) = k(5) = 7.$$

Therefore, the value of $k(h(6))$ is 7, which is option A.

14. Practice Test 2

Praxis Algebra I Practice Test

Total number of questions: 60

Time: <u>150 Minutes</u>

Calculator is permitted for Praxis Algebra I Test.

14.1 Practices

1) A student has two part-time jobs. Her combined work schedules consist of less than 50 hours per week.

   Which graph best represents the solution set for all possible combinations of x, the number of hours she worked at her first job, and y, the number of hours she worked at her second job, in one week?

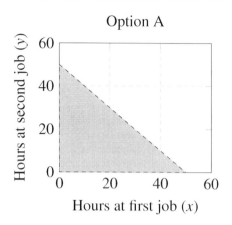
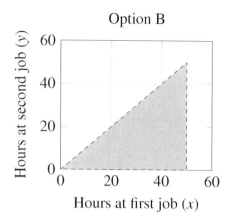
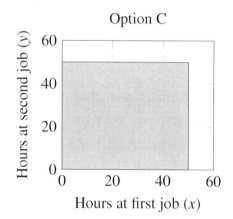
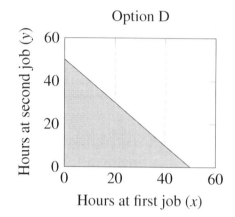

2) In the standard (x,y) coordinate plane, which of the following lines contains the points $(4,-10)$ and $(6,10)$?

☐ A. $y = 5x - 30$

☐ B. $y = -5x + 10$

☐ C. $y = 10x - 50$

☐ D. $y = -10x + 50$

3) Which of the following is equal to the expression below?

$$(3x - 4y)(5x + 2y).$$

☐ A. $15x^2 + 6xy - 8y^2$

☐ B. $15x^2 - 2y^2$

☐ C. $7x^2 + 10xy - 8y^2$

☐ D. $15x^2 - 14xy - 8y^2$

4) Which answer choice best represents the domain and range of the function $y = \sqrt{x+4}$?

14.1 Practices

☐ A. Domain: $x \geq -4$

Range: $y \geq 0$

☐ B. Domain: $-4 \leq x \leq 4$

Range: $y \geq 0$

☐ C. Domain: $y \geq 0$

Range: $x \geq -4$

☐ D. Domain: $x \leq -4$

Range: $y \geq 0$

5) What is the slope of a line that is perpendicular to the line $3x + 5y = 7$?

☐ A. $-\frac{3}{5}$

☐ B. $\frac{5}{3}$

☐ C. $-\frac{5}{3}$

☐ D. $\frac{3}{5}$

6) Tickets to a concert cost $10 for adults and $6 for children. A group of 10 people purchased tickets for $80. How many children's tickets did they buy?

☐ A. 2

☐ B. 4

☐ C. 5

☐ D. 6

7) Which graph best represents part of a quadratic function with a range of all real numbers less than -2?

Option A
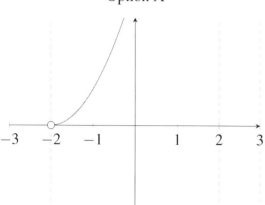

Option B
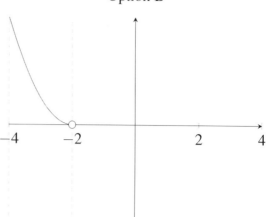

Option C
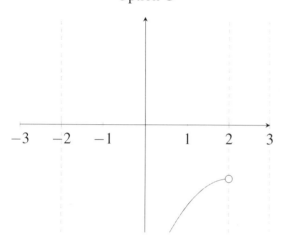

Option D
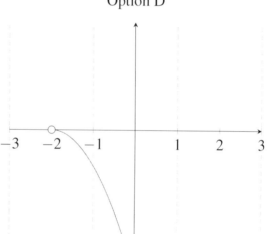

8) If the function is defined as $f(x) = ax^2 + 10$, where a is a constant, and $f(2) = 30$, what is the value of $f(3)$?

☐ A. 30

☐ B. 40

☐ C. 45

☐ D. 55

9) The graph of a line is shown on the grid. The coordinates of both points indicated on the graph of the line are integers.

What is the rate of change of y with respect to x for this line?

14.1 Practices

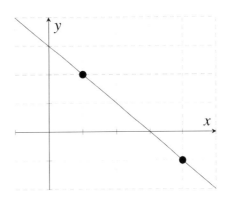

- [ ] A. $\frac{4}{3}$
- [ ] B. $-\frac{3}{4}$
- [ ] C. -2
- [ ] D. -1

10) What are the equation and slope of the line shown on the grid?

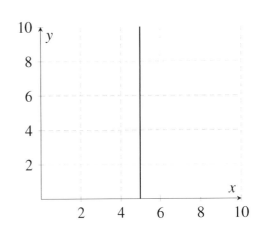

- [ ] A. $y = 5$; the slope is zero.
- [ ] B. $x = 5$; the slope is undefined.
- [ ] C. $y = 5$; the slope is 5.
- [ ] D. $x = 5$; the slope is 1.

11) In a sequence of numbers, $a_1 = 5$, $a_2 = 9$, $a_3 = 13$, $a_4 = 17$, and $a_5 = 21$. Based on this information, which equation can be used to find the nth term in the sequence, a_n?

- [ ] A. $a_n = n - 4$
- [ ] B. $a_n = n + 4$
- [ ] C. $a_n = 4n - 3$
- [ ] D. $a_n = 4n + 1$

12) A study tracks the growth of a plant over several weeks. The scatterplot and table show the number of weeks since planting and the height of the plant in centimeters. A linear function can be used to model this relationship.

Which function best models the data?

Weeks Since Planting (x)	Plant Height (cm) (y)
0	15.0
2	20.5
4	24.8
6	30.3
8	35.2

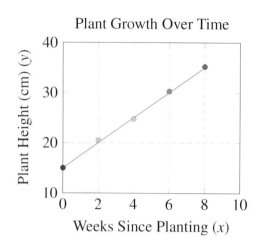

- ☐ A. $y = 2.5x + 15$
- ☐ B. $y = -3x + 20$
- ☐ C. $y = 20x - 3$
- ☐ D. $y = -3x + 25$

13) Which graph best represents a system of equations that has no solution?

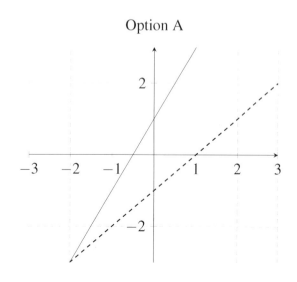

Option A

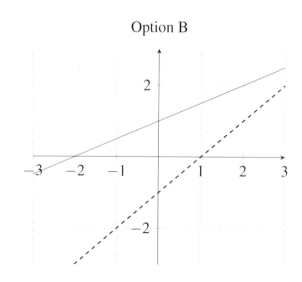

Option B

14.1 Practices

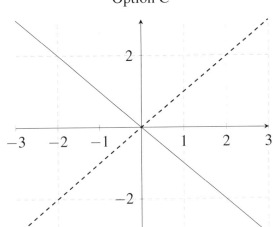

Option C

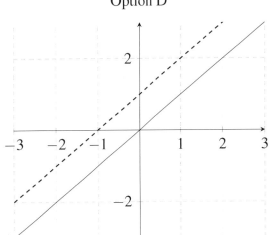

Option D

14) The expression $m^{-2}(m^3)^2$ is equivalent to m^y. What is the value of y? Write your answer in the box.

[ ]

15) What is the y-intercept of the line with the equation $2x - 5y = 10$?

- ☐ A. −4
- ☐ B. −2
- ☐ C. 2
- ☐ D. 5

16) The tables of ordered pairs represent some points on the graphs of lines L_1 and L_2. Which system of equations is represented by lines L_1 and L_2?

- ☐ A. $2x - 3y = 9$
       $x - 2y = -4$

- ☐ B. $3x - 2y = -9$
       $x - 2y = 4$

- ☐ C. $2x - 3y = 4$
       $x + y = -9$

- ☐ D. $x - 3y = 2$
       $2x + y = 4$

Line L_1

x	y
1	−1
2	0
5	2

Line L_2

x	y
−8	−1
−9	0
−10	1

17) Simplify $3x^3y^2 \left(2x^4y^3\right)^2 =$.

- ☐ A. $12x^7y^5$
- ☐ B. $12x^{11}y^8$
- ☐ C. $36x^{11}y^8$
- ☐ D. $36x^7y^5$

18) What are the zeroes of the function $f(x) = x^3 - 6x^2 + 9x$?

- ☐ A. 0
- ☐ B. 3
- ☐ C. 0, 3
- ☐ D. 0, −3

19) Which expression is equivalent to $0.00045 \times \left(2.5 \times 10^3\right)$?

- ☐ A. 1.125×10^{-1}
- ☐ B. 1.125
- ☐ C. 1.125×10
- ☐ D. 1.125×10^2

20) What is the positive solution to the equation $2(x+2)^2 = 18 - 11x$. Write your answer in the box.

☐

21) The height of a plant was measured each week for a twelve-week period. The graph shows a linear relationship between the height of the plant in inches and the number of weeks the plant was measured. Which statement best describes the y-intercept of the graph?

14.1 Practices

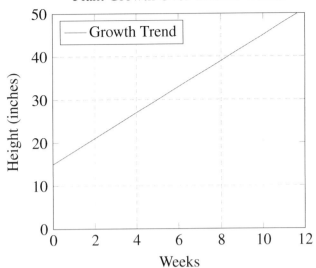

- ☐ A. The height of the plant was measured for 12 weeks.
- ☐ B. The maximum height was 48 inches.
- ☐ C. The height increased by 3 inches per week.
- ☐ D. The height of the plant at the beginning of the measurement period.

22) Find the axis of symmetry of the function $f(x) = -\frac{1}{4}(x+2)^2 + 5$.

- ☐ A. $y = 5$
- ☐ B. There is no axis of symmetry.
- ☐ C. $x = -2$
- ☐ D. $y = -\frac{1}{4}x + 5$

23) The initial value of a computer is $1,200. The value of the computer will decrease at a rate of 15% each year. Which graph best models this situation?

Option A

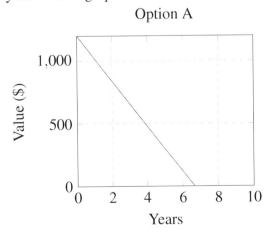

Option B

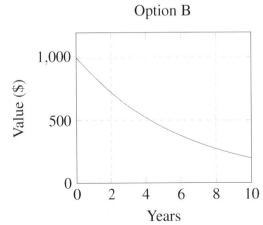

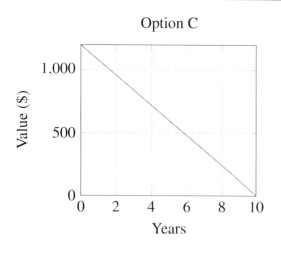

Option C

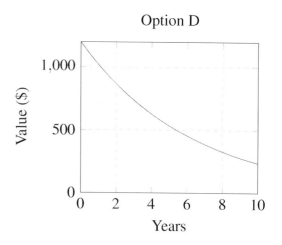

Option D

24) Which graph best represents the solution set of $3x + 4y \geq 12$?

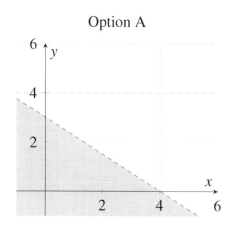

Option A

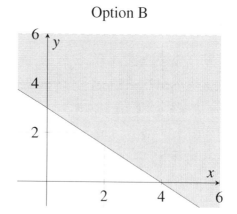

Option B

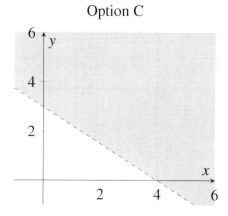

Option C

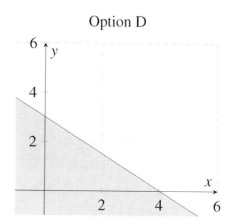

Option D

25) Which expression is equivalent to $m^2 - 14m + 45$?

☐ A. $(m+5)(m+9)$

☐ B. $(m-15)(m-3)$

☐ C. $(m+15)(m+3)$

14.1 Practices

☐ D. $(m-5)(m-9)$

26) The table shows the linear relationship between the revenue generated in thousand dollars by a bookshop and the number of books sold. What is the rate of change in revenue in thousand dollars with respect to the number of books sold in the shop?

☐ A. 0.04
☐ B. 0.06
☐ C. 0.02
☐ D. 0.08

Books Sold	Revenue (in thousand dollars)
50	2
100	4
150	6
200	8

27) Quadratic function f models the height in feet of a ball thrown into the air t seconds after it is thrown. The graph of the function is shown. What is the maximum value of the graph of the function? Write your answer in the box: ☐

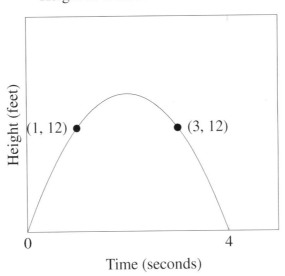

Height of a Thrown Ball Over Time

28) What is the value of z in the following system of equations?

$$\begin{cases} 4z + 3y = 6 \\ y = z \end{cases}$$

- [ ] A. $z = \frac{1}{2}$
- [ ] B. $z = \frac{3}{4}$
- [ ] C. $z = \frac{6}{7}$
- [ ] D. $z = \frac{5}{4}$

29) The graphs of linear functions h and k are displayed on the grid. Which function is best represented by the graph of k?

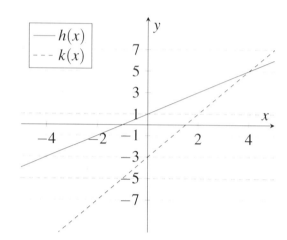

- [ ] A. $k(x) = 2h(x) + 5$
- [ ] B. $k(x) = \frac{1}{2}h(x) - 5$
- [ ] C. $k(x) = 2h(x) - 5$
- [ ] D. $k(x) = \frac{1}{2}h(x) + 5$

30) The equation $y^2 = 5y - 4$ has how many distinct real solutions?

- [ ] A. 0
- [ ] B. 1
- [ ] C. 2
- [ ] D. 3

31) A table represents some points on the graph of a linear function. Which equation represents the same relationship?

- [ ] A. $y - 5 = 3(x + 20)$
- [ ] B. $y - 20 = \frac{1}{3}(x + 5)$
- [ ] C. $y + 5 = \frac{1}{3}(x - 20)$
- [ ] D. $y + 20 = 3(x - 5)$

x	y
-5	20
0	$\frac{65}{3}$
15	$\frac{80}{3}$

32) Examine the graph of a linear equation provided below.

What are the coordinates of the x-intercept in this graph?

14.1 Practices

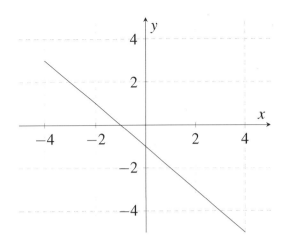

- A. $(0,-1)$
- B. $(0,-4)$
- C. $(-1,0)$
- D. $(-3,0)$

33) An exponential function is depicted on the provided grid. Which function most accurately represents the graph?

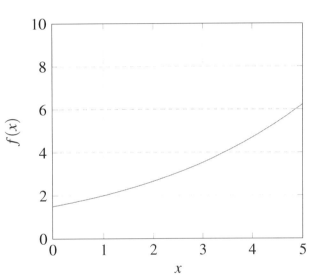

- A. $f(x) = 3(0.33)^x$
- B. $f(x) = 3(0.67)^x$
- C. $f(x) = 1.5(1.33)^x$
- D. $f(x) = 1.5(33)^x$

34) Given $x = 7$, calculate the value of y in the equation:

$$2y = \frac{3x^2}{5} - 5.$$

Write your answer in the box: ☐.

35) Determine the domain of the function $g(x) = -5x^2 + 36$.

- A. $(-\infty, 36]$
- B. $(-6, 6)$
- C. $[-3, 3]$
- D. $\mathbb{R}$

36) Which of the following is equal to $c^{\frac{4}{7}}$?

- A. $c^{\frac{7}{4}}$
- B. $\sqrt{c^{\frac{7}{4}}}$
- C. $\sqrt[7]{c^4}$
- D. $\sqrt[4]{c^7}$

37) On Sunday, John read P pages of a magazine each hour for 5 hours, and Lisa read Q pages of a magazine each hour for 2 hours. Which of the following represents the total number of magazine pages read by John and Lisa on Sunday?

- A. $7PQ$
- B. $10PQ$
- C. $5P + 2Q$
- D. $2P + 5Q$

38) A business analyzed the number of online transactions processed through their website since January 2019. The table shows the number of transactions in millions over time. This data can be modeled by a quadratic function.

Months Since Jan 2019	Transactions (millions)
0	5.0
6	5.0
12	41.0
18	113.0
24	221.0

Which function best models this data?

- A. $f(x) = 0.5x^2 - 3x + 5$
- B. $f(x) = -0.5x^2 + 3x - 5$
- C. $f(x) = 0.5x^2 + 3x + 5$
- D. $f(x) = x^2 - 6x + 10$

39) Which of the following represents the graph of the line with the equation $3x + 4y = 12$?

14.1 Practices

Option A

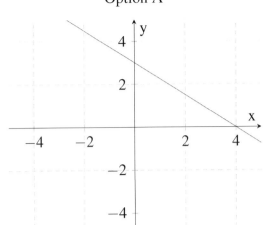

Option B

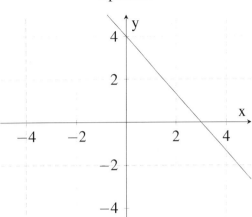

Option C

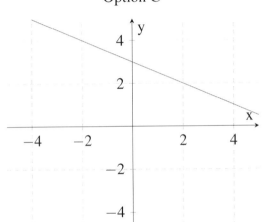

Option D

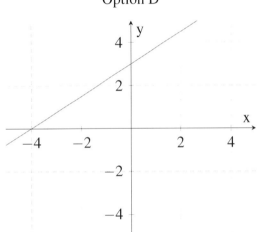

40) Determine the solution (x,y) for the following system of equations:

$$3x + 4y = 10$$
$$5x - 8y = -15$$

☐ A. $\left(\frac{5}{8}, \frac{9}{4}\right)$

☐ B. $\left(\frac{3}{8}, \frac{5}{8}\right)$

☐ C. $\left(\frac{5}{11}, \frac{95}{44}\right)$

☐ D. $\left(\frac{3}{11}, \frac{5}{44}\right)$

41) Find the solutions for the equation $4x^2 - 5x = 4 - x$.

☐ A. $x = 4$ and $x = 1$

- B. $x = 4$ and $x = -\frac{4}{5}$
- C. $x = \frac{2+\sqrt{5}}{4}$ and $x = \frac{2-\sqrt{5}}{4}$
- D. $x = \frac{1+\sqrt{5}}{2}$ and $x = \frac{1-\sqrt{5}}{2}$

42) The table details the linear relationship between the amount of water in a tank (in gallons) and the time in hours since filling began.

Time (hours)	Water (gallons)
0	0
2	50
4	100
6	150
8	200

Based on the table, what is the rate of change of the water level in the tank in gallons per hour? Write your answer in the box. ☐

43) The function $h(x)$ is described by a polynomial. Some values of x and $h(x)$ are listed in the following table. Which of the following must be a factor of $h(x)$?

- A. $x + 2$
- B. $x - 3$
- C. $x + 3$
- D. $x - 4$

x	$h(x)$
0	3
1	0
2	-1
3	0
4	3

44) Determine the value of $\frac{5d}{e}$ when $\frac{e}{d} = 3$.

- A. 0
- B. $\frac{3}{5}$
- C. $\frac{5}{3}$
- D. 3

45) Which of the following equations represents a graph that is a straight line?

14.1 Practices

- [ ] A. $y = 2x^2 + 8$
- [ ] B. $x^2 + 4y^2 = 4$
- [ ] C. $5x - 3y = 5x$
- [ ] D. $8x + 3xy = 9$

46) If $\frac{c-d}{d} = \frac{12}{15}$, then which of the following must be true?

- [ ] A. $\frac{c}{d} = \frac{12}{15}$
- [ ] B. $\frac{c}{d} = \frac{12}{27}$
- [ ] C. $\frac{c}{d} = \frac{27}{15}$
- [ ] D. $\frac{c}{d} = \frac{24}{12}$

47) Simplify the expression $3\sqrt{18} + 3\sqrt{3}$.

- [ ] A. $12\sqrt{5}$
- [ ] B. $6\sqrt{21}$
- [ ] C. $9\sqrt{2} + 3\sqrt{3}$
- [ ] D. $6\sqrt{3}$

48) Identify which table shows y as a function of x.

Table A:

x	y
1	2
2	4
2	4
4	8

Table B:

x	y
1	2
2	4
2	5
4	8

Table C:

x	y
1	2
2	4
4	6
4	8

Table D:

x	y
1	2
3	6
3	8
3	8

49) The graph of a quadratic function is displayed on a grid.

Which equation best represents the axis of symmetry for this graph?

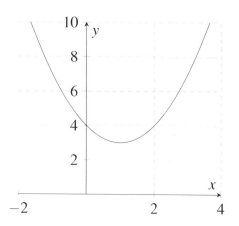

- A. $y = 8$
- B. $x = 1$
- C. $y = 4$
- D. $x = -2$

50) A tree increases in height at a constant rate. After six weeks, the tree is 36 cm tall. Which of the following functions represents the relationship between the height h of the tree and the number of weeks w since it was measured?

- A. $h(w) = 36w + 6$
- B. $h(w) = 6w + 36$
- C. $h(w) = 36w$
- D. $h(w) = 6w$

51) If $a ⊠ b = \sqrt{a^2 - b}$, what is the value of $7 ⊠ 21$?

- A. $2\sqrt{7}$
- B. $\sqrt{23}$
- C. $3\sqrt{14}$
- D. $\sqrt{17}$

52) Calculate the product of all possible values of y in the equation $|y - 8| = 5$.

- A. 2
- B. 8
- C. 11
- D. 39

53) The average of seven numbers is 32. If an eighth number, 50, is added, what is the new average? (Round your answer to the nearest hundredth.)

- A. 31

14.1 Practices

☐ B. 33.38

☐ C. 34.25

☐ D. 35.75

54) A credit union offers 3.75% simple interest on a savings account. If you deposit $15,000, how much interest will you earn in three years?

☐ A. $562.50

☐ B. $1,687.50

☐ C. $3,375

☐ D. $4,500

55) If $h(x) = 3x^3 + 4x^2 - x$ and $k(x) = -3$, what is the value of $h(k(x))$?

☐ A. -3

☐ B. -39

☐ C. 26

☐ D. -42

56) If 200% of a number is 100, then what is 60% of that number?

☐ A. 30

☐ B. 40

☐ C. 50

☐ D. 60

57) 10 students had an average score of 80. The remaining 8 students of the class had an average score of 90. What is approximately the mean (average) score of the entire class?

☐ A. 83.3

☐ B. 84.4

☐ C. 84.5

☐ D. 86.7

58) John orders a pack of notebooks for $4 per pack. A tax of 7.5% is added to the cost of the notebooks before a flat shipping fee of $5 rounds out the transaction. Which of the following represents the total cost of n packs of notebooks in dollars?

- A. $4.75n+5$
- B. $4n+4$
- C. $11.5n+4$
- D. $4.3n+5$

59) If $z-4=5$ and $3w+2=8$, what is the value of $zw+10$?

- A. 28
- B. 22
- C. 19
- D. 15

60) The following graph represents a polynomial function. What is the number of roots of this function?

- A. 0
- B. 1
- C. 2
- D. 3

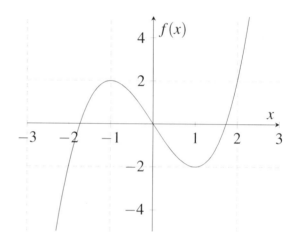

14.2　Answer Keys

1) A. Option A
2) C. $y = 10x - 50$
3) D. $15x^2 - 14xy - 8y^2$
4) A. Domain: $x \geq -4$
   Range: $y \geq 0$
5) B. $\frac{5}{3}$
6) C. 5
7) C. Option C
8) D. 55
9) D. -1
10) B. $x = 5$; the slope is undefined.
11) D. $a_n = 4n + 1$
12) A. $y = 2.5x + 15$
13) Option D.
14) 4
15) B. -2
16) C. $\begin{array}{l} 2x - 3y = 4 \\ x + y = -9 \end{array}$
17) B. $12x^{11}y^8$
18) C. 0, 3
19) B. 1.125
20) $\frac{1}{2}$
21) D.
22) C. $x = -2$
23) Option D.
24) Option B.
25) D. $(m-5)(m-9)$
26) A. 0.04

27) 16 feet
28) C. $z = \frac{6}{7}$
29) C. $k(x) = 2h(x) - 5$
30) C. 2
31) B. $y - 20 = \frac{1}{3}(x+5)$
32) C. $(-1, 0)$
33) C. $f(x) = 1.5(1.33)^x$
34) 12.2
35) D. $\mathbb{R}$
36) C. $\sqrt[7]{c^4}$
37) C. $5P + 2Q$
38) A. $f(x) = 0.5x^2 - 3x + 5$
39) A. Option A
40) C. $\left(\frac{5}{11}, \frac{95}{44}\right)$
41) D. $x = \frac{1+\sqrt{5}}{2}$ and $x = \frac{1-\sqrt{5}}{2}$
42) 25 gallons per hour
43) B. $x - 3$
44) C. $\frac{5}{3}$
45) C. $5x - 3y = 5x$
46) C. $\frac{c}{d} = \frac{27}{15}$
47) C. $9\sqrt{2} + 3\sqrt{3}$
48) Table A
49) B. $x = 1$
50) D. $h(w) = 6w$
51) A. $2\sqrt{7}$
52) D. 39
53) C. 34.25
54) B. $1,687.50

55) D. −42

56) A. 30

57) B. 84.4

58) D. $4.3n+5$

59) A. 28

60) D. 3

14.3 Answers with Explanation

1) The correct graph must represent all combinations of x and y where the total hours worked $(x+y)$ are less than 50. In Option A, the shaded area below the dashed line $y = 50 - x$ shows all points where the sum of x and y is less than 50, correctly representing the constraints of the problem. The other options either represent incorrect relationships or do not correctly depict the total hours constraint. In option D, the line is solid which is incorrect.

2) To find the equation of the line that passes through $(4, -10)$ and $(6, 10)$, we first calculate the slope:

$$\text{Slope} = \frac{10 - (-10)}{6 - 4} = \frac{20}{2} = 10.$$

Using the point-slope form of a line, $y - y_1 = m(x - x_1)$, and the point $(4, -10)$:

$$y - (-10) = 10(x - 4) \Rightarrow y + 10 = 10x - 40 \Rightarrow y = 10x - 40 - 10 \Rightarrow y = 10x - 50.$$

Therefore, the equation of the line is $y = 10x - 50$, which corresponds to option C.

3) Multiplying the polynomials $(3x - 4y)(5x + 2y)$ using the distributive property (FOIL method):

$$3x \times 5x = 15x^2, \quad 3x \times 2y = 6xy, \quad -4y \times 5x = -20xy, \quad -4y \times 2y = -8y^2.$$

Adding these together:

$$15x^2 + 6xy - 20xy - 8y^2 = 15x^2 - 14xy - 8y^2.$$

So, the correct answer is option D.

4) For the function $y = \sqrt{x+4}$, the domain and range can be determined as follows:

- Domain: Since the square root function is defined for non-negative numbers, $x + 4 \geq 0$. Therefore, $x \geq -4$.

- Range: The square root function always yields non-negative values. Thus, $y \geq 0$.

Therefore, the correct answer is option A, with the domain $x \geq -4$ and the range $y \geq 0$.

5) The slope of the line $3x + 5y = 7$ can be found by rearranging it into slope-intercept form $y = mx + b$:

$$5y = -3x + 7,$$

$$y = -\frac{3}{5}x + \frac{7}{5}.$$

The slope of this line is $-\frac{3}{5}$. The slope of a line perpendicular to this line is the negative reciprocal of $-\frac{3}{5}$, which is $\frac{5}{3}$.

Therefore, the correct answer is option B, $\frac{5}{3}$.

6) Let the number of children's tickets be c and the number of adult tickets be $10 - c$. The total cost is $80. Setting up the equation:

$$10(10 - c) + 6c = 80.$$

Solving for c:

$$100 - 10c + 6c = 80 \Rightarrow -4c = -20 \Rightarrow c = 5.$$

Therefore, the number of children's tickets bought is 5, which is option C.

7) The correct graph must represent a quadratic function with a range of all real numbers less than -2.

In Options A and B, the range is $y > 0$.

In Option D, the range is $y < 0$.

In Option C, the graph correctly models the condition that the function's range is all real numbers less than -2.

8) First, find the value of a using $f(2) = 30$:

$$30 = a(2)^2 + 10 \Rightarrow 30 = 4a + 10 \Rightarrow 4a = 20 \Rightarrow a = 5.$$

Now, find $f(3)$ using $a = 5$:

$$f(3) = 5(3)^2 + 10,$$

$$f(3) = 5 \times 9 + 10 = 45 + 10 = 55.$$

Therefore, the value of $f(3)$ is 55, which is option D.

14.3 Answers with Explanation

9) The rate of change (slope) of a line is calculated using two points on the line. For the points $(1,2)$ and $(4,-1)$, the slope is calculated as the difference in y-coordinates divided by the difference in x-coordinates:

$$\text{Slope} = \frac{-1-2}{4-1} = -1.$$

This slope indicates that for each unit increase in x, y decreases by 1 unit.

Therefore, the correct answer is option D.

10) The graph shows a vertical line crossing the x-axis at $x = 5$. For a vertical line, the x-coordinate is constant while the y-coordinate changes. This line is described by the equation $x = 5$.

The slope of a vertical line is undefined because the change in y is not zero while the change in x is zero, leading to a division by zero when calculating the slope. Thus, the correct answer is option B.

11) To find the nth term of the sequence, observe the pattern: each term increases by 4. The first term is 5, the second is $9 = (5+4)$, the third is $13 = (9+4)$, and so on. The formula for the nth term can be found by noticing that each term is four times the term number plus 1:

$$a_n = 4n + 1.$$

Therefore, the correct equation for the nth term in this sequence is option D, $a_n = 4n + 1$.

12) The scatter plot and table show a linear relationship between the number of weeks since planting and the height of the plant. The function $y = 2.5x + 15$ approximately models this relationship. In this function, the slope of 2.5 indicates that the plant grows by 2.5 cm each week, and the y-intercept of 15 represents the initial height of the plant at the time of planting (0 weeks). The other options either have incorrect slopes or y-intercepts that do not align with the data presented.

13) A system of linear equations has no solution when the lines represented by the equations are parallel and never intersect. Among the options, Option D is the only one that shows two parallel lines. Since these lines never meet, there is no point that satisfies both equations simultaneously, indicating that the system has no solution. The other options depict intersecting lines, suggesting systems with one solution.

14) To simplify the expression $m^{-2}(m^3)^2$ and find y:

$$m^{-2}(m^3)^2 = m^{-2} \cdot m^6 = m^{-2+6} = m^4.$$

Therefore, the expression is equivalent to m^4, and the value of y is 4.

15) The y-intercept of a line occurs where $x = 0$. Substituting $x = 0$ into the equation $2x - 5y = 10$:

$$2(0) - 5y = 10 \Rightarrow -5y = 10 \Rightarrow y = -2.$$

Therefore, the y-intercept is -2, which is option B.

16) The tables for lines L_1 and L_2 show points that satisfy the equations $2x - 3y = 4$ and $x + y = -9$, respectively. These equations correspond to Option C. The points in the table for L_1 fit the equation $2x - 3y = 4$, and the points in the table for L_2 fit the equation $x + y = -9$. This makes Option C the correct choice, as it represents the system of equations depicted by the tables.

17) To simplify the expression $3x^3y^2 \left(2x^4y^3\right)^2$:

$$3x^3y^2 \cdot (2x^4y^3)^2 = 3x^3y^2 \cdot 4x^8y^6 = 12x^{11}y^8.$$

Therefore, the simplified expression is $12x^{11}y^8$, which is option B.

18) To find the zeroes of $f(x) = x^3 - 6x^2 + 9x$, factor out the common term:

$$f(x) = x(x^2 - 6x + 9).$$

The quadratic can be factored as:

$$f(x) = x(x-3)^2.$$

Setting each factor to zero gives the zeroes:

$$x = 0, \; x - 3 = 0 \Rightarrow x = 3.$$

14.3 Answers with Explanation

Therefore, the zeroes are $0, 3$, which is option C.

19) To simplify $0.00045 \times (2.5 \times 10^3)$:

$$0.00045 \times 2.5 \times 10^3 = 4.5 \times 10^{-4} \times 2.5 \times 10^3 = 11.25 \times 10^{-4+3} = 11.25 \times 10^{-1}.$$

Converting 11.25×10^{-1} back to decimal form:

$$11.25 \times \frac{1}{10} = 1.125.$$

Therefore, the equivalent expression is 1.125, which is option B.

20) To solve the equation $2(x+2)^2 = 18 - 11x$:

$$2(x+2)^2 + 11x = 18 \Rightarrow 2(x^2 + 4x + 4) + 11x = 18$$
$$\Rightarrow 2x^2 + 19x + 8 = 18$$
$$\Rightarrow 2x^2 + 19x - 10 = 0.$$

By factoring we have:
$$(2x-1)(x+10) = 0.$$

The positive solution:
$$2x - 1 = 0 \Rightarrow x = \frac{1}{2}.$$

Therefore, the positive solution of the equation is $x = \frac{1}{2}$.

21) The y-intercept of the graph represents the height of the plant at the beginning of the measurement period (week 0), which is 15 inches in this case. It does not reflect the duration of the measurement, the maximum height, or the final height. The slope of the line, which is 3, indicates that the height of the plant increased by 3 inches per week. Therefore, the best description of the y-intercept is provided by Option D.

22) The axis of symmetry of a quadratic function $f(x) = a(x-h)^2 + k$ is given by the line $x = h$. For the function $f(x) = -\frac{1}{4}(x+2)^2 + 5$, the form matches $a(x+2)^2 + k$, where $h = -2$.

Therefore, the axis of symmetry is $x = -2$, which is option C.

23) The correct graph for this scenario is Option D, which shows an exponential decay starting at $1,200 and decreasing by 15% each year. The graph depicts the value as continuously decreasing, but the rate of decrease slows down over time, which is characteristic of exponential decay. The value approaches but never reaches zero, accurately representing the depreciation of the computer over time.

24) The inequality $3x + 4y \geq 12$ is represented by a line and the region above it. Option B correctly uses a solid line to indicate that points on the line are included in the solution set (the equality part of the inequality). The shaded area above the line in Option B shows the region where $3x + 4y$ is greater than 12. This graph accurately represents the solution set of the given inequality.

25) To factorize $m^2 - 14m + 45$, look for two numbers that multiply to 45 and add up to -14. The numbers -5 and -9 fit this criterion:

$$m^2 - 14m + 45 = (m-5)(m-9).$$

Therefore, the equivalent expression is $(m-5)(m-9)$, which is option D.

26) The rate of change in revenue can be determined by calculating the slope of the line representing the relationship between the number of books sold and the revenue generated. From the table, the slope is calculated as the change in revenue divided by the change in the number of books sold. The change in revenue is $6,000 (from $2,000 to $8,000), and the change in books sold is 150 (from 50 to 200 books). The slope is therefore:

$$\frac{\$6 \text{ thousand dollars}}{150 \text{ books}} = \$0.04 \text{ thousand dollars per book.}$$

This represents an increase in revenue of $0.04 thousand dollars for each additional book sold, making Option A the correct answer.

27) By solving the system of equations derived from the given points (1, 12), (3, 12), and (4, 0) for the quadratic function $f(x) = ax^2 + bx + c$ (with $c = 0$ as the graph passes through (0, 0)), we find that the maximum value of the function f at $x = 2$ is 16. The graph of quadratic function $f(x) = ax^2 + bx + c$ passes through $(0,0)$. Thus, $f(0) = 0$ and hence $c = 0$.

14.3 Answers with Explanation

We see that the graph also passes through points $(1,12)$, $(3,12)$, and $(4,0)$. Therefore, we have:

$$f(1) = 12 \Rightarrow a(1)^2 + b(1) = 12 \Rightarrow a+b = 12$$
$$f(3) = 12 \Rightarrow a(3)^2 + b(3) = 12 \Rightarrow 9a+3b = 12.$$

By solving the above system of equations, we find $a = -4$ and $b = 16$. Hence:

$$f(x) = -4x^2 + 16x.$$

To find the maximum value of f, which occurs at the vertex of the parabola, we evaluate $f(2)$ as the vertex lies midway between $x = 1$ and $x = 3$ (the symmetry of the parabola):

$$\text{Maximum value} = f(2) = -4(2)^2 + 16(2) = -16 + 32 = 16.$$

28) Substituting $y = z$ into the first equation:

$$4z + 3z = 6 \Rightarrow 7z = 6 \Rightarrow z = \frac{6}{7}.$$

Therefore, the value of z is $\frac{6}{7}$, which is option C.

29) We can consider two points of each graph and calculate the slpoe of each line. In the graph, h is a linear function with a slope of 1. The slpoe of function k is 2. Thus,

$$k(x) = 2h(x) + a,$$

where a is a constant. Since $h(4) = k(4) = 5$, then we get:

$$5 = 2(5) + a \Rightarrow a = -5.$$

Therefore, $k(x) = 2h(x) - 5$, making option C correct.

30) Rewrite the equation in standard quadratic form:

$$y^2 - 5y + 4 = 0.$$

Factor this equation, we obtain:

$$(y-4)(y-1) = 0.$$

The solutions are $y = 4$ and $y = 1$, which are distinct real solutions.

Therefore, the equation has 2 distinct real solutions, which is option C.

31) The table provides the points $(-5, 20)$, $(0, \frac{65}{3})$, and $(15, \frac{80}{3})$ that lie on the graph of a linear function. These points align with the equation $y = \frac{1}{3}x + \frac{65}{3}$, which is the simplified form of Option B. The relationship between x and y in these points indicates a slope of $\frac{1}{3}$ and a y-intercept of $\frac{65}{3}$. This matches the equation in Option B, making it the correct choice for representing the same relationship.

32) The x-intercept of a graph is the point where the line crosses the x-axis, which occurs where the y-coordinate is zero. In the provided graph, the line intersects the x-axis at the point $(-1, 0)$. Therefore, the coordinates of the x-intercept are $(-1, 0)$, which corresponds to Option C.

33) The provided graph shows an exponential increase, which is consistent with the function in Option C, $f(x) = 1.5(1.33)^x$. This function starts with a base value of 1.5 and grows at a rate of 1.33 per x increment, indicating an exponential growth pattern. The other options either decrease exponentially (Options A and B) or increase at a significantly different rate (Option D), which does not match the graph's pattern. Therefore, Option C is the most accurate representation of the graph.

34) Substituting $x = 7$ into the equation:

$$2y = \frac{3 \times 7^2}{5} - 5 = \frac{3 \times 49}{5} - 5 = \frac{147}{5} - 5 = 29.4 - 5 = 24.4.$$

Solving for y:

$$y = \frac{24.4}{2} = 12.2.$$

35) Since $g(x) = -5x^2 + 36$ is a quadratic function, it is defined for all real numbers. Therefore, the domain

14.3 Answers with Explanation

is all real numbers, $\mathbb{R}$.

36) $c^{\frac{4}{7}}$ is equivalent to the 7-th root of c^4, which is $\sqrt[7]{c^4}$.

37) John reads $5P$ pages in total (5 hours times P pages per hour), and Lisa reads $2Q$ pages in total (2 hours times Q pages per hour). The total number of pages read is $5P + 2Q$.

38) The correct equation is identified by examining the trend of the data in the table and comparing it to the equations of the given options. The table shows a quadratic trend with an increasing rate of transactions over time. Option A, $f(x) = 0.5x^2 - 3x + 5$, closely aligns with this trend, as it correctly models the increase in the number of transactions over the months. The coefficients and constants in this equation match the pattern observed in the data, making it the best choice for representing this relationship.

39) To find the correct graph for the line $3x + 4y = 12$, the equation is first rearranged into slope-intercept form, which is $y = mx + b$. After rearrangement, the equation becomes $y = -\frac{3}{4}x + 3$. Option A correctly shows this line, with a slope of $-\frac{3}{4}$ and a y-intercept of 3. The other options either have incorrect slopes or incorrect y-intercepts, making them incorrect representations of the line.

40) Solving the system of equations:

From $3x + 4y = 10$:

$$4y = 10 - 3x \Rightarrow y = \frac{10 - 3x}{4}.$$

Substitute y in $5x - 8y = -15$:

$$5x - 8\left(\frac{10 - 3x}{4}\right) = -15 \Rightarrow 5x - 2(10 - 3x) = -15 \Rightarrow 5x + 6x - 20 = -15 \Rightarrow 11x = 5.$$

Thus, $x = \frac{5}{11}$. Now, we can find y:

$$y = \frac{10 - 3\left(\frac{5}{11}\right)}{4} = \frac{1}{4}\left(\frac{95}{11}\right) = \frac{95}{44}.$$

Therefore, the point $\left(\frac{5}{11}, \frac{95}{44}\right)$ is the solution, option C.

41) Rearranging and simplifying the equation:

$$4x^2 - 5x - 4 + x = 0 \Rightarrow 4x^2 - 4x - 4 = 0 \Rightarrow x^2 - x - 1 = 0.$$

Using the quadratic formula:

$$x = \frac{-b \pm \sqrt{b^2 - 4ac}}{2a}$$

$$x = \frac{1 \pm \sqrt{(-1)^2 - 4 \cdot 1 \cdot (-1)}}{2 \cdot 1}$$

$$x = \frac{1 \pm \sqrt{1 + 4}}{2}$$

$$x = \frac{1 \pm \sqrt{5}}{2}.$$

Thus, the solutions are $x = \frac{1+\sqrt{5}}{2}$ and $x = \frac{1-\sqrt{5}}{2}$.

42) The rate of change of the water level can be determined by calculating the slope of the line representing the relationship between the water level and time. The slope is the change in the water level divided by the change in time. From the table, the water level increases by 200 gallons over 8 hours, resulting in a rate of change of $\frac{200 \text{ gallons}}{8 \text{ hours}} = 25$ gallons per hour. This indicates that the water level in the tank increases by 25 gallons for each hour.

43) To determine the correct factors of the polynomial $h(x)$, we look for values of x such that $h(x) = 0$. In the given table, $h(x)$ equals 0 at $x = 1$ and $x = 3$. This indicates that $x - 1$ and $x - 3$ must be factors of $h(x)$, as the polynomial function equals zero when x is 1 or 3. Therefore, Option B, $x - 3$, is the correct answer.

44) Given $\frac{e}{d} = 3$, we can rearrange it to $e = 3d$. Substituting this into $\frac{5d}{e}$:

$$\frac{5d}{3d} = \frac{5}{3}.$$

Therefore, $\frac{5d}{e} = \frac{5}{3}$.

45) The equation $5x - 3y = 5x$ simplifies to $-3y = 0$ or $y = 0$, which is a horizontal line. Thus, it represents a graph that is a straight line.

14.3 Answers with Explanation

46) Given $\frac{c-d}{d} = \frac{12}{15}$, we can rewrite it as:

$$\frac{c}{d} - 1 = \frac{12}{15}.$$

Adding 1 to both sides:

$$\frac{c}{d} = \frac{12}{15} + 1 = \frac{12}{15} + \frac{15}{15} = \frac{27}{15}.$$

Therefore, $\frac{c}{d} = \frac{27}{15}$.

47) Simplify the expression:

$$3\sqrt{18} + 3\sqrt{3} = 3\sqrt{9 \times 2} + 3\sqrt{3} = 3 \times 3\sqrt{2} + 3\sqrt{3} = 9\sqrt{2} + 3\sqrt{3}.$$

Since $\sqrt{2}$ and $\sqrt{3}$ cannot be simplified further and are not like terms, the expression is already in its simplest form.

48) A function is defined as a relationship where each input (x) has exactly one output (y). In Option A, every x-value corresponds to one unique y-value, which meets the definition of a function. The other options have duplicate x-values with different y-values, which is not characteristic of a function. Therefore, Option A correctly shows y as a function of x.

49) The axis of symmetry of a quadratic function is a vertical line that passes through the vertex of the parabola. In this graph, the vertex appears to be near $x = 1$. Therefore, the axis of symmetry would be the line $x = 1$. Option B, $x = 1$, is the correct answer as it best represents the axis of symmetry of the graph.

50) The tree grows 36 cm over 6 weeks, so the rate of growth per week is $\frac{36\ cm}{6\ weeks} = 6\ cm/week$. Therefore, the height after w weeks can be represented by the function $h(w) = 6w$.

51) Using the custom operation ⊠:

$$7 \boxtimes 21 = \sqrt{7^2 - 21} = \sqrt{49 - 21} = \sqrt{28} = \sqrt{4 \times 7} = 2\sqrt{7}.$$

Therefore, the correct answer is option A.

52) Solve the equation $|y-8|=5$ for y:

Case 1: $y-8=5$ gives $y=13$.

Case 2: $y-8=-5$ gives $y=3$.

The product of the two solutions is $13 \times 3 = 39$.

53) The total sum of the first seven numbers is $7 \times 32 = 224$. Adding the eighth number, 50, gives a new total of $224+50=274$. The new average for eight numbers is $\frac{274}{8} = 34.25$.

54) Simple interest can be calculated using the formula $I = P \times r \times t$, where I is the interest, P is the principal amount, r is the rate of interest, and t is the time in years.

For a deposit of $\$15,000$ at a rate of 3.75% for 3 years:

$$I = \$15,000 \times 3.75\% \times 3 = \$15,000 \times 0.0375 \times 3 = \$1,687.50.$$

55) First, substitute $k(x) = -3$ into $h(x)$:

$$h(k(x)) = h(-3) = 3(-3)^3 + 4(-3)^2 - (-3).$$

Calculate the value:

$$= 3(-27) + 4(9) + 3 = -81 + 36 + 3 = -42.$$

Therefore, the correct answer is option D.

56) First, find the number. Since 200% of the number is 100:

$$\frac{200}{100} \times \text{number} = 100 \Rightarrow \text{number} = \frac{100}{2} = 50.$$

Next, calculate 60% of 50:

$$\frac{60}{100} \times 50 = 30.$$

Therefore, the number is 30, making option A correct.

14.3 Answers with Explanation

57) Calculate the total score for both groups:

10 students × 80 score/student = 800 total score.

8 students × 90 score/student = 720 total score.

The total number of students is $10+8=18$, and the total score is $800+720=1520$. The average score is:

$$\frac{1520}{18} \approx 84.44.$$

Rounded to the nearest tenth, this is approximately 84.4.

58) The cost of n packs is $4n$. With 7.5% tax, the total becomes $4n \times 1.075 = 4.3n$. Adding the flat shipping fee of $5 gives the total cost as $4.3n + 5$.

59) First, solve for z and w: $z-4=5$ gives $z=9$. $3w+2=8$ gives $3w=6$ and $w=2$.

Now, calculate $zw+10$: $9 \times 2 + 10 = 18 + 10 = 28$, which is option A.

60) The roots of a polynomial function are the x-values where the graph intersects the x-axis. In the provided graph, the polynomial function intersects the x-axis at three distinct points. Therefore, it has three roots, making Option D, 3 roots, the correct answer.

Author's Final Note

I hope you enjoyed this book as much as I enjoyed writing it. I have tried to make it as easy to understand as possible. I have also tried to make it fun. I hope I have succeeded. If you have any suggestions for improvement, please let me know. I would love to hear from you.

The accuracy of examples and practice is very important to me. We have done our best. But I also expect that I have made some minor errors. Constant improvement is the name of the game. If you find any errors, please let me know. I will fix them in the next edition.

Your learning journey does not end here. I have written a series of books to help you learn math. Make sure you browse through them. I especially recommend workbooks and practice tests to help you prepare for your exams.

I also enjoy reading your reviews. If you have a moment, please leave a review on Amazon. It will help other students find this book.

If you have any questions or comments, please feel free to contact me at drNazari@effortlessmath.com.

And one last thing: Remember to use online resources for additional help. I recommend using the resources on `https://effortlessmath.com`. There are many great videos on YouTube.

Good luck with your studies!

Dr. Abolfazl Nazari

Made in the USA
Columbia, SC
15 October 2024